Springer Undergraduate Mathematics Series

The Springer Undergraduate Mathematics Series (SUMS) is a series designed for undergraduates in mathematics and the sciences worldwide. From core foundational material to final year topics, SUMS books take a fresh and modern approach. Textual explanations are supported by a wealth of examples, problems and fully-worked solutions, with particular attention paid to universal areas of difficulty. These practical and concise texts are designed for a one- or two-semester course but the self-study approach makes them ideal for independent use.

More information about this series at http://www.springer.com/series/3423

Viorel Barbu

Differential Equations

Springer

Viorel Barbu
Department of Mathematics
Alexandru Ioan Cuza University
Iaşi
Romania

ISSN 1615-2085 ISSN 2197-4144 (electronic)
Springer Undergraduate Mathematics Series
ISBN 978-3-319-45260-9 ISBN 978-3-319-45261-6 (eBook)
DOI 10.1007/978-3-319-45261-6

Library of Congress Control Number: 2016954710

Mathematics Subject Classification (2010): 34A12, 34A30, 34A60, 34D20, 35F05

Translation from the Romanian language edition: Ecuaţii diferenţiale by Viorel Barbu © 1985 Junimea, Iaşi, Romania. All Rights Reserved.

Printed on acid-free paper

This Springer imprint is published by Springer Nature
The registered company is Springer International Publishing AG
The registered company address is: Gewerbestrasse 11, 6330 Cham, Switzerland

Preface

The present book is devoted to the study of differential equations. It is well known that to write a monograph or a textbook in a classical subject is a difficult enterprise requiring a rigorous selection of topics and exposition techniques. Part of mathematical analysis, the theory of differential equations is a fundamental discipline that carries a considerable weight in the professional development of mathematicians, physicists and engineers and, to a lesser extent, that of biologists and economists. Through its topics and investigative techniques, this discipline has a broad scope, touching diverse areas such as topology, functional analysis, mechanics, mathematical and theoretical physics, and differential geometry.

Although this book is structured around the main problems and results of the theory of ordinary differential equations, it contains several more recent results which have had a significant impact on research in this area. In fact, even when studying classical problems we have opted for techniques that highlight the functional methods and which are also applicable to evolution equations in infinite-dimensional spaces, thus smoothing the way towards a deeper understanding of the modern methods in the theory of partial differential equations.

We wrote this work bearing in mind the fact that differential equations represent, in truth, a branch of applied mathematics and that the vast majority of such equations have their origin in the mathematical modelling of phenomena in nature or society. We tried to offer the reader a large sample of such examples and applications in the hope that they will stimulate his interest and will provide a strong motivation for the study of this theory. Each chapter ends with a number of exercises and problems, theoretical or applied and of varying difficulty.

Since this book is not a treatise, and since we wanted to keep it within a reasonable size, we had to conscientiously omit several problems and subjects that are typically found in classical texts devoted to ordinary differential equations such as periodic systems of differential equations, Carathéodory solutions, delay-differential equations, differential equations on manifolds, and Sturm–Liouville problems. The willing reader can take up these topics at a later stage. In fact, the list of references contains the titles of several monographs or college textbooks that can substitute these omissions and complement the present book.

In closing, I would like to thank my colleagues Dr. Nicolae Luca and Dr. Gheorghe Morosanu, who read the manuscript and made pertinent comments that I took into account.

Iaşi, Romania
December 1982

Viorel Barbu

Acknowledgements

This is the English translation of the Romanian edition of the book published with Junimea Publishing House, Iaşi, 1985. The English translation was done by Prof. Liviu Nicolaescu from Notre Dame University (USA). I want to thank him for this and also for some improvements he included in this new version. Many thanks are due also to Dr. Gabriela Marinoschi and Professor Cătălin Lefter, who corrected errors and suggested several improvements.

Iaşi, Romania
May 2016

Contents

Frequently Used Notation

- $\mathbb{Z}$—the set of integers.
- $\mathbb{Z}_{\geq k}$—the set integers $\geq k$.
- $\mathbb{R}$—the set of real numbers
- $(a, b) := \{x \in \mathbb{R};\ \ a < x < b\}$.
- $\mathbb{R}^+ := (0, \infty)$, $\overline{\mathbb{R}}^+ := [0, \infty)$
- $\mathbb{C}$—the set of complex numbers.
- $\boldsymbol{i} := \sqrt{-1}$.
- $(-, -) :=$ the Euclidean inner product in $\mathbb{R}^n$

$$(\boldsymbol{x}, \boldsymbol{y}) := \sum_{j=1}^{n} \boldsymbol{x}_j \boldsymbol{y}_j, \quad x = (x_1, \ldots, x_n), \quad y = (y_1, \ldots, y_n).$$

- $\|-\|_{\mathbf{e}} :=$ the Euclidean norm on $\mathbb{R}^n$,

$$\|\boldsymbol{x}\|_{\mathbf{e}} = \sqrt{(\boldsymbol{x}, \boldsymbol{x})} = \sqrt{\boldsymbol{x}_1^2 + \cdots + \boldsymbol{x}_n^2}, \ x = (x_1, \ldots, x_n) \in \mathbb{R}^n.$$

- 2^S—the collection of all subsets of the set S.

Chapter 1
Introduction

This chapter is devoted to the concept of a solution to the Cauchy problem for various classes of differential equations and to the description of several classical differential equations that can be explicitly solved. Additionally, a substantial part is devoted to the discussion of some physical problems that lead to differential equations. Much of the material is standard and can be found in many books; notably [1, 4, 6, 9].

1.1 The Concept of a Differential Equation

Loosely speaking, a *differential equation* is an equation which describes a relationship between an unknown function, depending on one or several variables, and its derivatives up to a certain order. The highest order of the derivatives of the unknown function that are involved in this equation is called the *order* of the differential equation. If the unknown function depends on several variables, then the equation is called a *partial differential equation*, or *PDE*. If the unknown function depends on a single variable, the equation is called an *ordinary differential equation*, or *ODE*. In this book, we will be mostly interested in ODEs. We investigate first-order PDEs in Chap. 5.

A first-order ODE has the general form

$$F(t, x, x') = 0, \tag{1.1}$$

where t is the argument of the unknown function $x = x(t)$, $x'(t) = \frac{dx}{dt}$ is its derivative, and F is a real-valued function defined on a domain of the space $\mathbb{R}^3$.

We define a *solution* of (1.1) on the interval $I = (a, b)$ of the real axis to be a continuously differentiable function $x : I \to \mathbb{R}$ that satisfies Eq. (1.1) on I, that is,

$$F\big(t, x(t), x'(t)\big) = 0, \quad \forall t \in I.$$

V. Barbu, *Differential Equations*, Springer Undergraduate Mathematics Series,
DOI 10.1007/978-3-319-45261-6_1

When I is an interval of the form $[a, b]$, $[a, b)$ or $(a, b]$, the concept of a solution on I is defined similarly.

In certain situations, the implicit function theorem allows us to reduce (1.1) to an equation of the form

$$x' = f(t, x), \tag{1.2}$$

where $f : \Omega \to \mathbb{R}$, with Ω an open subset of $\mathbb{R}^2$. In the sequel, we will investigate exclusively equations in the form (1.2), henceforth referred to as *normal form*.

From a geometric viewpoint, a solution of (1.1) is a curve in the (t, x)-plane, having at each point a tangent line that varies continuously with the point. Such a curve is called an *integral curve* of Eq. (1.1). In general, the set of solutions of (1.1) is infinite and we will (loosely) call it the *general solution* of (1.1). We can specify a solution of (1.1) by imposing certain conditions. The most frequently used is the *initial condition* or the *Cauchy condition*

$$x(t_0) = x_0, \tag{1.3}$$

where $t_0 \in I$ and $x_0 \in \mathbb{R}$ are a priori given and are called *initial values*.

The *Cauchy problem* associated with (1.1) asks to find a solution $x = x(t)$ of (1.1) satisfying the initial condition (1.3). Geometrically, the Cauchy problem amounts to finding an integral curve of (1.1) that passes through a given point $(t_0, x_0) \in \mathbb{R}^2$.

Phenomena in nature or society usually have a pronounced dynamical character. They are, in fact, processes that evolve in time according to their own laws. The movement of a body along a trajectory, a chemical reaction, an electrical circuit, biological or social groups are the simplest examples. The investigation of such a process amounts to following the evolution of a finite number of parameters that characterize the corresponding process or system.

Mathematically, such a group of parameters describes the *state* of the system or the process and represents a group of functions that depend on time. For example, the movement of a point in space is completely determined by its coordinates $x(t) = (x_1(t), x_2(t), x_3(t))$. These parameters characterize the state of this system.

The state of a biological population is naturally characterized by the number of individuals it is made of. The state of a chemical reaction could be, depending on the context, the temperature or the concentration of one or several of the substances that participate in the reaction.

The state is rarely described by an explicit function of time. In most cases, the state of a system is a solution of an equation governing the corresponding phenomenon according to its own laws. Modelling a dynamical phenomenon boils down to discovering this equation, which very often is a differential equation. The initial condition (1.3) signifies that the state of the system at time t_0 is prescribed. This leads us to expect that this initial condition uniquely determines the solution of equation (1.1) or (1.2). In other words, given the pair $(t_0, x_0) \in I \times \mathbb{R}$ we expect that (1.2) has only one solution satisfying (1.3). This corresponds to the deterministic point of view in the natural sciences according to which the evolution of a process is uniquely determined by its initial state. We will see that under sufficiently mild assumptions on the function f this is indeed true. The precise statement and the proof of this result, known as the *existence and uniqueness theorem*, will be given in the next chapter.

The above discussion extends naturally to first-order *differential systems* of the form

$$x_i' = f_i(t, x_1, \dots, x_n), \quad i = 1, \dots, n, \quad t \in I, \tag{1.4}$$

where $f_1, \dots, f_n$ are functions defined on an open subset of $\mathbb{R}^{n+1}$. By a solution of system (1.4) we understand a collection of continuously differentiable functions $\{x_1(t), \dots, x_n(t)\}$ on the interval $I \subset \mathbb{R}$ that satisfy (1.4) on this interval, that is,

$$x_i'(t) = f_i\big(t, x_1(t), \dots, x_n(t)\big), \quad i = 1, \dots, n, \quad t \in I, \tag{1.5}$$

$$x_i(t_0) = x_i^0, \quad i = 1, \dots, n, \tag{1.6}$$

where $t_0 \in I$ and $(x_1^0, \dots, x_n^0)$ is a given point in $\mathbb{R}^n$. Just as in the scalar case, we will refer to (1.5)–(1.6) as the *Cauchy problem* associated with (1.4), named after the French mathematician *A.L. Cauchy* (1789–1857), who first defined this problem and proved the existence and uniqueness of its solutions.

From a geometric point of view, a solution of system (1.4) is a curve in the space $\mathbb{R}^n$. In many situations or phenomena modelled by differential systems of type (1.4), the collection $\big(x_1(t), \dots, x_n(t)\big)$ represents the coordinates of the state of a system at time t, and thus the trajectory $t \mapsto \big(x_1(t), \dots, x_n(t)\big)$ describes the evolution of that particular system. For this reason, the solutions of a differential system are often called the *trajectories* of the system.

Consider now ordinary differential equations of order n, that is, having the form

$$F\big(t, x, x', \dots, x^{(n)}\big) = 0, \tag{1.7}$$

where F is a given function. Assuming it is possible to solve for $x^{(n)}$, we can reduce the above equation to its *normal form*

$$x^{(n)} = f\big(t, x, \dots, x^{(n-1)}\big). \tag{1.8}$$

By a solution of (1.8) on the interval I we understand a function of class C^n on I (that is, a function n-times differentiable on I with continuous derivatives up to order n) that satisfies (1.8) at every $t \in I$. The Cauchy problem associated with (1.8) asks to find a solution of (1.8) that satisfies the conditions

$$x(t_0) = x_0^0, \quad x'(t_0) = x_1^0, \dots, x^{(n-1)}(t_0) = x_{n-1}^0, \tag{1.9}$$

where $t_0 \in I$ and $x_0^0, x_1^0, \dots, x_{n-1}^0$ are given.

Via a simple transformation we can reduce Eq. (1.8) to a system of type (1.4). To this end, we introduce the new unknown functions $x_1, \dots, x_n$ using the unknown function x by setting

$$x_1 := x, \quad x_2 := x', \dots, x_n := x^{(n-1)}. \tag{1.10}$$

In this notation, Eq. (1.8) becomes the differential system

$$\begin{aligned} x_1' &= x_2 \\ x_2' &= x_3 \\ \vdots \;&\vdots\; \vdots \\ x_n' &= f(t, x_1, \dots, x_n). \end{aligned} \tag{1.11}$$

Conversely, any solution of (1.11) defines via (1.10) a solution of (1.8). The change of variables (1.10) transforms the initial conditions (1.1) into

$$x_i(t_0) = x_{i-1}^0, \quad i = 1, \dots, n.$$

The above procedure can also be used to transform differential systems of order n (that is, differential systems containing derivatives up to order n) into differential systems of order 1. There are more general ordinary differential equations where the unknown functions appears in the equation with different arguments. This is the case for differential equations of the form

$$x'(t) = f\big(t, x(t), x(t-h)\big), \quad t \in I, \tag{1.12}$$

where h is a positive constant. Such an equation is called *delay-differential* equation. It models certain phenomena that have "memory", in other words, physical processes in which the present state is determined by the state at a certain period of time in the past. It is a proven fact that many phenomena in nature and society fall into this category. For example in physics, the hysteresis and visco-elasticity phenomena display such behaviors. It is also true that phenomena displaying "memory" are sometimes described by more complicated functional equations, such as the integro-differential equations of Volterra type:

$$x'(t) = f_0\big(t, x(t)\big) + \int_a^t K\big(t, s, x(s)\big)ds, \tag{1.13}$$

named after the mathematician *V. Volterra* (1860–1940) who introduced and studied them for the first time. The integral term in (1.13) incorporates the "history" of the phenomenon. If we take into account that the derivative $\frac{dx}{dt}$ at t is computed using the values of x at t and at infinitesimally close moments $t - \varepsilon$, we can even say that the differential equation (1.1) or (1.2) "has memory", only in this case we are talking of "short-term-memory" phenomena.

There exist processes or phenomena whose states cannot be determined by finitely many functions of time. For example, the concentration of a substance in a chemical reaction or the amplitude of a vibrating elastic system are functions that depend both on time and on the location in the space where the process is taking place. Such processes are called *distributed* and are typically modelled by partial differential equations.

1.2 Elementary Methods of Solving Differential Equations

The first differential equation was solved as soon as differential and integral calculus was invented in the 17th century by *I. Newton* (1642–1727) and *G.W. Leibniz* (1646–1716). We are speaking, of course, of the problem

$$x'(t) = f(t), \quad t \in I, \tag{1.14}$$

where f is a continuous function. As we know, its solutions are given by

$$x(t) = x_0 + \int_{t_0}^{t} f(s)ds, \quad t \in I.$$

The 18th century and a large part of the 19th century were dominated by the efforts of mathematicians such as *J. Bernoulli* (1667–1748), *L. Euler* (1707–1783), *J. Lagrange* (1736–1813) and many others to construct explicit solutions of differential equations. In other words, they tried to express the general solution of a differential equation as elementary functions or as primitives (antiderivatives) of such functions. Without trying to force an analogy, these efforts can be compared to those of algebrists who were trying to solve by radicals higher degree polynomial equations. Whereas in algebra *E. Galois* (1811–1832) completely solved the problem of solvability by radicals of algebraic equations, in analysis the problem of explicit solvability of differential equations has lost its importance and interest due to the introduction of qualitative and numerical methods of investigating differential equations. Even when we are interested only in approximating solutions of a differential equation, or understanding their qualitative features, having an explicit formula, whenever possible, is always welcome.

1.2.1 Equations with Separable Variables

These are equations of the form

$$\frac{dx}{dt} = f(t)g(x), \quad t \subset I = (a, b), \tag{1.15}$$

where f is a continuous function on (a, b) and g is a continuous function on a, possibly unbounded, interval (x_1, x_2), $g \neq 0$. Note that we can rewrite equation (1.15) as

$$\frac{dx}{g(x)} = f(t)dt.$$

Integrating from t_0 to t, where t_0 is an arbitrary point in I, we deduce that

$$\int_{x_0}^{x(t)} \frac{d\tau}{g(\tau)} = \int_{t_0}^{t} f(s)ds. \tag{1.16}$$

We set

$$G(x) := \int_{x_0}^{x} \frac{d\tau}{g(\tau)}. \tag{1.17}$$

The function G is obviously continuous and monotone on the interval (x_1, x_2). It is thus invertible and its inverse has the same properties. We can rewrite (1.16) as

$$x(t) = G^{-1}\left(\int_{t_0}^{t} f(s)ds\right), \quad t \in I. \tag{1.18}$$

We have thus obtained a formula describing the solution of (1.15) satisfying the Cauchy condition $x(t_0) = x_0$. Conversely, the function x given by equality (1.18) is continuously differentiable on I and its derivative satisfies

$$x'(t) = \frac{f(t)}{G'(x)} = f(t)g(x).$$

In other words, x is a solution of (1.15). Of course, $x(t)$ is only defined for those values of t such that $\int_{t_0}^{t} f(s)ds$ lies in the range of G.

By way of illustration, consider the ODE

$$x' = (2 - x)\tan t, \quad t \in \left(0, \frac{\pi}{2}\right),$$

where $x \in (-\infty, 2) \cup x \in (2, \infty)$. Arguing as in the general case, we rewrite this equation in the form

$$\frac{dx}{2 - x} = \tan t.$$

Integrating

$$\int_{x_0}^{x} \frac{\theta}{2 - \theta} d\theta = \int_{t_0}^{t} \tan s \, ds, \quad t_0, t \in \left(0, \frac{\pi}{2}\right), \quad x(t_0) = x_0,$$

we deduce that

$$\ln \frac{|x(t) - 2|}{|x_0 - 2|} = -\ln \frac{|\cos t|}{|\cos t_0|} = \ln \frac{\cos t_0}{\cos t}.$$

If we set $C := |x_0 - 2| \cos t_0$, we deduce that the general solution is given by

$$x(t) = \frac{C}{\cos t} + 2, \quad t \in \left(0, \frac{\pi}{2}\right),$$

where C is an arbitrary constant.

1.2.2 Homogeneous Equations

Consider the differential equation

$$x' = h(x/t), \tag{1.19}$$

where h is a continuous function defined on an interval (h_1, h_2). We will assume that $h(r) \neq r$ for any $r \in (h_1, h_2)$. Equation (1.19) is called a *homogeneous* differential equation. It can be solved by introducing a new unknown function u defined by the equality $x = tu$. The new function u satisfies the separable differential equation

$$tu' = h(u) - u,$$

which can be solved by the method described in Sect. 1.2.1. We have to mention that many first-order ODEs can be reduced by simple substitutions to separated or homogeneous differential equations. Consider, for example, the differential equation

$$x' = \frac{at + bx + c}{a_1t + b_1x + c_1},$$

where a, b, c and a_1, b_1, c_1 are constants. This equation can be reduced to a homogeneous equation of the form

$$\frac{dy}{ds} = \frac{as + by}{a_1s + b_1y}$$

by making the change of variables

$$s := t - t_0, \quad y := x - x_0,$$

where (t_0, x_0) is a solution of the linear algebraic system

$$at_0 + bx_0 + c = a_1t_0 + b_1x_0 + c_0 = 0.$$

1.2.3 First-Order Linear Differential Equations

Consider the differential equation

$$x' = a(t)x + b(t), \tag{1.20}$$

where a and b are continuous functions on the, possibly unbounded, interval (t_1, t_2). To solve (1.20), we multiply both sides of this equation by

$$\exp\left(-\int_{t_0}^{t} a(s)ds\right),$$

where t_0 is some point in (t_1, t_2). We obtain

$$\frac{d}{dt}\left(\exp\left(-\int_{t_0}^{t} a(s)ds\right)x(t)\right) = b(t)\exp\left(-\int_{t_0}^{t} a(s)ds\right).$$

Hence, the general solution of (1.20) is given by

$$x(t) = \exp\left(\int_{t_0}^{t} a(s)ds\right)\left(x_0 + \int_{t_0}^{t} b(s)\exp\left(-\int_{t_0}^{s} a(\tau)d\tau\right)ds\right), \tag{1.21}$$

where x_0 is an arbitrary real number. Conversely, differentiating (1.21), we deduce that the function x defined by this equality is the solution of (1.20) satisfying the Cauchy condition $x(t_0) = x_0$.

Consider now the differential equation

$$x' = a(t)x + b(t)x^{\alpha}, \tag{1.22}$$

where α is a real number not equal to 0 or 1. Equation (1.22) is called a *Bernoulli type* equation and can be reduced to a linear equation using the substitution $y = x^{1-\alpha}$.

1.2.4 Exact Differential Equations

Consider the equation

$$x' = \frac{g(t,x)}{h(t,x)}, \tag{1.23}$$

where g and h are continuous functions defined on an open set $\Omega \subset \mathbb{R}^2$. We assume additionally that $h \neq 0$ in Ω and that the expression $hdx - gdt$ is an *exact differential*. This means that there exists a differentiable function $F \in C^1(\Omega)$ such that $dF = hdx - gdt$, that is,

$$\frac{\partial F}{\partial x}(t,x) = h(t,x), \quad \frac{\partial F}{\partial t}(t,x) = -g(t,x), \quad \forall (t,x) \in \Omega. \tag{1.24}$$

Equation (1.23) becomes

$$dF\big(t, x(t)\big) = 0.$$

Hence every solution x of (1.23) satisfies the equality

$$F\big(t, x(t)\big) = C, \tag{1.25}$$

where C is an arbitrary constant. Conversely, for any constant C, equality (1.25) defines via the implicit function theorem (recall that $\frac{\partial F}{\partial x} = h \neq 0$ in Ω) a unique function $x = x(t)$ defined on some interval (t_1, t_2) and which is a solution of (1.23).

1.2.5 Riccati Equations

Named after *J. Riccati* (1676–1754), these equations have the general form

$$x' = a(t)x + b(t)x^2 + c, \quad t \in I, \tag{1.26}$$

where a, b, c are continuous functions on the interval I. In general, Eq. (1.26) is not explicitly solvable but it enjoys several interesting properties which we will dwell upon later. Here we only want to mention that if we know a particular solution $\varphi(t)$ of (1.26), then, using the substitution $y = x - \varphi$, we can reduce Eq. (1.26) to a Bernoulli type equation in y. We leave to the reader the task of verifying this fact.

1.2.6 Lagrange Equations

These are equations of the form

$$x = t\varphi(x') + \psi(x'), \tag{1.27}$$

where φ and ψ are two continuously differentiable functions defined on a certain interval of the real axis such that $\varphi(p) \neq p$, $\forall p$. Assuming that x is a solution of (1.27) on the interval $I \subset \mathbb{R}$, we deduce after differentiating that

$$x' = \varphi(x') + t\varphi'(x')x'' + \psi'(x')x'', \tag{1.28}$$

where $x'' = \frac{d^2x}{dt^2}$. We denote by p the function x' and we observe that (1.28) implies that

$$\frac{dt}{dp} = \frac{\varphi'(p)}{p - \varphi(p)}t + \frac{\psi'(p)}{p - \varphi(p)}. \tag{1.29}$$

We can interpret (1.29) as a linear ODE with unknown t, viewed as a function of p. Solving this equation by using formula (1.21), we obtain for t an expression of the form

$$t = A(p, C), \tag{1.30}$$

where C is an arbitrary constant. Using this in (1.27), we deduce that

$$x = A(p, C)\varphi(p) + \psi(p). \tag{1.31}$$

If we interpret p as a parameter, equalities (1.30) and (1.31) define a parametrization of the curve in the (t, x)-plane described by the graph of the function x. In other words, the above method leads to a parametric representation of the solution of (1.27).

1.2.7 Clairaut Equations

Named after *A.C. Clairaut* (1713–1765), these equations correspond to the degenerate case $\varphi(p) \equiv p$ of (1.27) and they have the form

$$x = tx' + \psi(x'). \tag{1.32}$$

Differentiating the above equality, we deduce that

$$x' = tx'' + x' + \psi'(x')x''$$

and thus

$$x''\big(t + \psi'(x')\big) = 0. \tag{1.33}$$

We distinguish two types of solutions. The first type is defined by the equation $x'' = 0$. Hence

$$x = C_1 t + C_2, \tag{1.34}$$

where C_1 and C_2 are arbitrary constants. Using (1.34) in (1.32), we see that C_1 and C_2 are not independent but are related by the equality

$$C_2 = \psi(C_1).$$

Therefore,

$$x = C_1 t + \psi(C_1), \tag{1.35}$$

where C_1 is an arbitrary constant. This is the *general solution* of the Clairaut equation.

A second type of solution is obtained from (1.33),

$$t + \psi'(x') = 0. \tag{1.36}$$

Proceeding as in the case of Lagrange equations, we set $p := x'$ and we obtain from (1.36) and (1.32) the parametric equations

$$t = -\psi'(p), \qquad x = -\psi'(p)p + \psi(p) \tag{1.37}$$

that describe a function called the *singular solution* of the Clairaut equation (1.32). It is not difficult to see that the solution (1.37) does not belong to the family of solutions (1.35). Geometrically, the curve defined by (1.37) is the envelope of the family of lines described by (1.35).

1.2.8 Higher Order ODEs Explicitly Solvable

We discuss here several classes of higher order ODEs that can be reduced to lower order ODEs using elementary changes of variables.

One first example is supplied by equations of the form

$$F\big(t, x^{(k)}, x^{(k+1)}, \dots, x^{(n)}\big) = 0, \tag{1.38}$$

where $0 < k < n$. Using the substitution $y := x^{(k)}$, we reduce (1.38) to

$$F\big(t, y, \dots, y^{(n-k)}\big) = 0.$$

If we can determine y from the above equation, the unknown x can be obtained from the equation

$$x^{(k)} = y \tag{1.39}$$

via repeated integration and we obtain

$$x(t) = \sum_{j=0}^{k-1} C_j \frac{(t-a)^j}{j!} + \frac{1}{k!}\int_a^t (t-s)^k y(s) ds. \tag{1.40}$$

Consider now ODEs of the form

$$F(x, x', \dots, x^{(n)}) = 0. \tag{1.41}$$

We set $p := x'$ and we think of p as our new unknown function depending on the independent variable x. We now have the obvious equalities

$$x'' = \frac{dp}{dt} = \frac{dp}{dx}p, \;\; x''' = \frac{d}{dx}\Big(\frac{dp}{dx}p\Big)p.$$

In general, $x^{(k)}$ can be expressed as a nonlinear function of $p, \frac{dp}{dx}, \dots, \frac{d^{k-1}p}{dx^{k-1}}$. When we replace $x^{(k)}$ by this expression in (1.41), we obtain an ODE of order $(n-1)$ in p and x. In particular, the second-order ODE

$$F(x, x', x'') = 0 \tag{1.42}$$

reduces via the above substitution to a first-order equation

$$F(x, p, \dot{p}) = 0, \quad \dot{p} := \frac{dp}{dx}.$$

For example, the *Van der Pol* equation (*B. Van der Pol* (1889–1959))

$$x'' + (x^2 - 1)x' + x = 0,$$

reduces to the first-order ODE

$$\dot{p} + xp^{-1} = 1 - x^2.$$

1.3 Mathematical Models Described by Differential Equations

In this section, we will present certain classical or more recent examples of ODEs or systems of ODEs that model certain physical phenomena. Naturally, we are speaking of dynamical models. Many more examples can be found in the book Braun (1978) from our list of references.

1.3.1 Radioactive Decay

It has been experimentally verified that the rate of radioactive decay is proportional to the number of atoms in the decaying radioactive substance. Thus, if $x(t)$ denotes the quantity of radioactive substance that is available at time t, then the rate of decay $x'(t)$ is proportional to $x(t)$, that is,

$$-x'(t) = \alpha x(t), \tag{1.43}$$

where α is a positive constant that depends on the nature of the radioactive substance. In other words, $x(t)$ satisfies a linear equation of type (1.20) and thus

$$x(t) = x_0 \exp\big(-\alpha(t - t_0)\big), \quad t \in \mathbb{R}. \tag{1.44}$$

Usually, the rate of decay is measured by the so-called "*half-life*", that is, the time it takes for the substance to radiate half of its mass. Equality (1.44) implies easily that the half-life, denoted by T, is given by the formula

$$T = \frac{\ln 2}{\alpha}.$$

In particular, the well-known radiocarbon dating method is based on this equation.

1.3.2 Population Growth Models

If $p(t)$ is the population of a certain species at time t and $d(t, p)$ is the difference between the birth rate and mortality rate, assume that the population is isolated, that is, there are no emigrations or immigrations. Then the rate of growth of the population will be proportional to $d(t, p)$. A simplified population growth model assumes that $d(t, p)$ is proportional to the population p. In other words, p satisfies the differential equation

$$p' = \alpha p, \quad \alpha \equiv constant. \tag{1.45}$$

The solution of (1.45) is, therefore,

$$p(t) = p_0 e^{\alpha(t-t_0)}. \tag{1.46}$$

This leads to the Malthusian law of population growth.

A more realistic model was proposed by the Dutch biologist *P. Verhulst* (1804–1849) in 1837. In Verhulst's model, the difference $d(t, p)$ is assumed to be $\alpha p - \beta p^2$, where β is a positive constant, a lot smaller than α. This nonlincar growth model, which takes into account the interactions between the individual of the species and, more precisely, the inhibitive effect of crowding, leads to the ODE

$$p' = \alpha p - \beta p^2. \tag{1.47}$$

It is interesting that the above equation also models the spread of technological innovations.

Equation (1.47) is a separable ODE and, following the general strategy for dealing with such equations, we obtain the solution

$$p(t) = \frac{\alpha p_0}{\beta p_0 + (\alpha - \beta p_0)\exp(-\alpha(t - t_0))}, \tag{1.48}$$

where (t_0, p_0) are the initial conditions.

A more complex biological system is that in which two species S_1 and S_2 share the same habitat so that the individuals of the species S_2, the predators, feed exclusively on the individuals of the species S_1, the prey. If we denote by $N_1(t)$ and $N_2(t)$ the number of individuals in, respectively, the first and second species at time t, then a mathematical model of the above biological system is described by the *Lotka–Volterra* system of ODEs

$$N_1'(t) = aN_1 - bN_1N_2, \qquad N_2'(t) = -cN_2 + dN_1N_2, \tag{1.49}$$

where a, b, c, d are positive constants. This system, often called the "*predator-prey*" system, is the inspiration for more sophisticated models of the above problem that lead to more complex equations that were brought to scientists' attention by V. Volterra in a classical monograph published in 1931.

1.3.3 Epidemic Models

We present here a classical mathematical model for epidemic spread proposed in 1927 by W.O. Kermac and A.G. McKendrick.

Consider a population consisting of n individuals and an infectious disease that spreads through direct contact. We assume that the infected individuals will either be isolated, or they will become immune after recovering from the disease. Therefore, at a given moment of time t, the population is comprised of three categories of individuals: *uninfected* individuals, *infected individuals roaming freely*, and *isolated* individuals. Let the sizes of these categories be $x(t)$, $y(t)$ and $z(t)$, respectively. We will assume that the infection rate $-x'(t)$ is proportional to the number xy which represents the number of possible contacts between uninfected and infected individuals. Also, we assume that the infected individuals are being isolated at a rate proportional to their number y. Therefore, the equations governing this process are

$$x' = -\beta xy, \quad y' = \beta xy - \gamma y, \tag{1.50}$$

$$x + y + z = n. \tag{1.51}$$

Using (1.50), we deduce that

$$\frac{x'}{y'} = -\frac{\beta x}{\beta x - \gamma} \Rightarrow \frac{dy}{dx} = \frac{\gamma - \beta x}{\beta x}.$$

Integrating, we get

$$y(x) = y_0 + x_0 - x + \frac{\gamma}{\beta} \ln \frac{x}{x_0},$$

where x_0, y_0 are the initial values of x and y. Invoking (1.51), we can now also express z as a function of x. To find x, y, z as functions of t it suffices to substitute y as described above into the first equation of (1.50) and then integrate the newly obtained equation.

1.3.4 The Harmonic Oscillator

Consider the motion of a particle of mass m that is moving along the x-axis under the influence of an elastic force directed towards the origin. We denote by $x(t)$ the position of the particle on the x-axis at time t. Newton's second law of dynamics implies that

$$mx'' = F. \tag{1.52}$$

On the other hand, F being an elastic force, it is of the form $F = -\omega^2 x$. We conclude that the motion of the particle is described by the second-order ODE

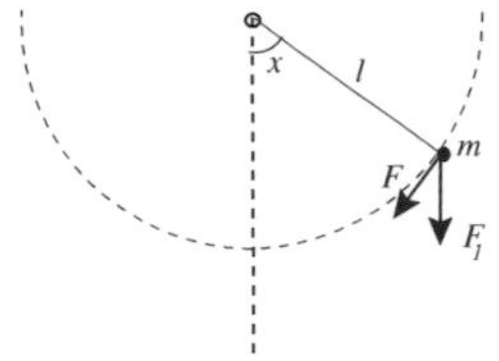

Fig. 1.1 A pendulum

$$mx'' + \omega^2 x = 0. \tag{1.53}$$

A more sophisticated model of this motion is one in which we allow for the presence of a resistance force of the form $-bx'$ and of an additional external force $f(t)$ acting on the particle. We obtain the differential equation

$$mx'' + bx' + \omega^2 x = f. \tag{1.54}$$

If the force F is not elastic but depends on the position x according to a more complicated law, Eqs. (1.53) and (1.54) will be nonlinear

$$mx'' + bx' + F(x) = f.$$

Consider, for example, the motion of a pendulum of mass m with rigid arm of length l that moves in a vertical plane; see Fig. 1.1. We denote by $x(t)$ the angle between the arm and the vertical direction. The motion is due to the gravitational force $F_1 = mg$, where g is the gravitational acceleration. This force has an active component F, tangent to the circular trajectory, and of size $F = -mg \sin x$. Invoking Newton's second law again, we deduce that this force must equal mlx'', the product of the mass and the acceleration. Therefore, the equation of motion is

$$lx'' + g \sin x = 0. \tag{1.55}$$

1.3.5 The Motion of a Particle in a Conservative Field

Consider a particle of mass m that moves under the influence of a force field $\boldsymbol{F} : \mathbb{R}^3 \to \mathbb{R}^3$. We denote by F_1, F_2, F_3 the components of the vector field $\boldsymbol{F}$ and by $x_1(t), x_2(t)$ and $x_3(t)$ the coordinates of the location of the particle at time t. Newton's second law then implies

$$mx_i''(t) = F_i\big(x_1(t), x_2(t), x_3(t)\big), \quad i = 1, 2, 3. \tag{1.56}$$

The vector field $\boldsymbol{F}$ is called *conservative* if there exists a C^1 function $U : \mathbb{R}^3 \to \mathbb{R}$ such that

$$F = -\nabla U \Longleftrightarrow F_i = -\frac{\partial U}{\partial x_i}, \quad \forall i = 1, 2, 3. \tag{1.57}$$

The function U is called the *potential* energy of the field $\boldsymbol{F}$. An elementary computation involving (1.56) yields the equality

$$\frac{d}{dt}\left(\frac{m}{2}\sum_{i=1}^{3}\left|x_i'(t)\right|^2 + U\big(x_1(t), x_2(t), x_3(t)\big)\right) = 0.$$

In other words, along the trajectory of the system, the energy

$$E = \frac{m}{2}\sum_{i=1}^{3}\left|x_i'(t)\right|^2 + U(x_1(t), x_2(t), x_3(t))$$

is conserved.

The harmonic oscillator discussed earlier corresponds to a linear force field $\boldsymbol{F}(\boldsymbol{x}) = -\omega^2\boldsymbol{x}$, $\boldsymbol{x} = (x_1, x_2, x_3)$. A vector field $\boldsymbol{F} : \mathbb{R}^3 \to \mathbb{R}^3$ is called *central* if it has the form

$$\boldsymbol{F}(\boldsymbol{x}) = f\big(\|\boldsymbol{x}\|_{\mathbf{e}}\big)\boldsymbol{x}, \tag{1.58}$$

where $f : \mathbb{R} \to \mathbb{R}$ is a given function and $\|\boldsymbol{x}\|_{\mathbf{e}}$ is the Euclidean norm

$$\|\boldsymbol{x}\|_{\mathbf{e}} := \sqrt{x_1^2 + x_2^2 + x_3^2}.$$

Without a doubt, the most important example of a central field is the gravitational force field. If the Sun is placed at the origin of the space $\mathbb{R}^3$, then it generates a gravitational force field of the form

$$\boldsymbol{F}(\boldsymbol{x}) = -\frac{gmM}{\|\boldsymbol{x}\|_{\mathbf{e}}^3}\boldsymbol{x}, \tag{1.59}$$

where M is the mass of the Sun, and m is the mass of a planet situated in this field. We are dealing with a conservative field with potential $U(\boldsymbol{x}) = -\frac{gmM}{\|\boldsymbol{x}\|_{\mathbf{e}}}$.

1.3.6 The Schrödinger Equation

In a potential field U, the steady states of the one-dimensional motion of a particle of mass m are described by the second-order ODE

$$\psi'' + \frac{2m}{\hbar^2}\big(E - U(x)\big)\psi = 0. \tag{1.60}$$

Fig. 1.2 An RLC circuit

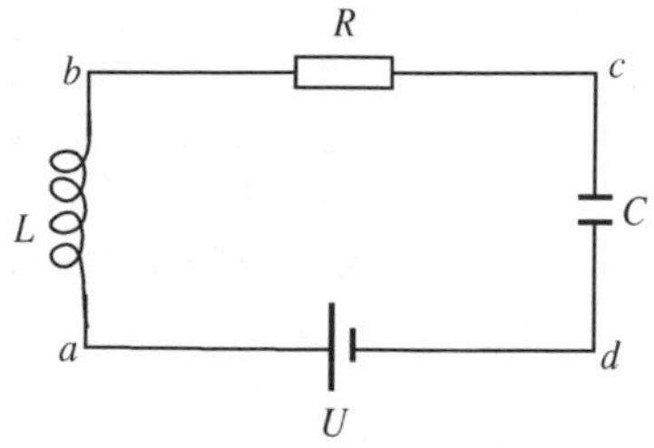

Equation (1.60), first described in 1926, is called the *Schrödinger equation* after its discoverer (the physicist *E. Schrödinger* (1887–1961)). It is the fundamental equation of non-relativistic quantum mechanics. In (1.60) $\hbar$ is *Planck's constant* and E is the energy level of the steady state. The function ψ is called the *wave function* of the particle. This means that for any $\alpha < \beta$ the probability that the particle can be detected in the interval $[\alpha, \beta]$ is $\int_\alpha^\beta |\psi(x)|^2 dx$.

1.3.7 Oscillatory Electrical Circuits

Consider an electrical circuit made of a coil with inductance L, a resistor with resistance R, a capacitor or condenser with capacitance C, and a source of electricity which produces a potential difference (or voltage) U (Fig. 1.2).

If we denote by $I(t)$ the intensity of the electric current, and by U_{ab}, U_{bc}, U_{cd}, U_{da} the potential drops across the segments ab, bc, cd and da respectively, we obtain from the basic laws of electricity the following equations

$$U_{ab}(t) = LI'(t), \quad U_{bc}(t) = RI(t), \quad CU'_{cd}(t) = I(t), \quad U_{da} = U.$$

Kirchoff's second law implies that $U_{ab} + U_{bc} + U_{cd} + U_{da} = 0$. Hence

$$LI''(t) + RI'(t) + \frac{1}{C}I(t) = f(t), \tag{1.61}$$

where $f(t) = U'(t)$. It is remarkable that Eq. (1.54) of the harmonic oscillator is formally identical to Eq. (1.61) of the oscillatory electrical circuit.

1.3.8 Solitons

Consider the function $u = u(x, t)$ depending on the variables t and x. We denote by u_{tt} and u_{xx} the second-order partial derivatives of u with respect to t and x, respectively. Suppose that u satisfies the partial differential equation

$$u_{tt} - C^2 u_{xx} = 0, \quad t \geq 0, \quad x \in \mathbb{R}, \tag{1.62}$$

where C is a real constant. The variable x is the "position" variable while t is the "time" variable. Equation (1.62) is known in mathematical physics as the *wave equation*. It describes, among other things, the vibration of an elastic string. The function $u(x, t)$ is the amplitude of the vibration at location x on the string and at time t.

A *traveling wave* of (1.62) is a solution of the form

$$u(x, t) = \varphi(x + \mu t), \tag{1.63}$$

where μ is a constant and φ is a one-variable C^2-function. Using u defined by (1.63) in Eq. (1.62), we deduce that $\mu = \pm C$. In other words, Eq. (1.62) admits two families of solutions of type (1.63), and *D'Alembert's principle* states that the general solution of (1.62) is

$$U(x, t) = \varphi(x + Ct) + \psi(x - Ct), \tag{1.64}$$

where $\varphi, \psi : \mathbb{R} \to \mathbb{R}$ are arbitrary C^2-functions. In general, a solution of type (1.64) for a given equation is called a *soliton*. The importance of solitons in mathematical physics stems from the fact that their general features tend to survive interactions. We want to emphasize that the solitons are found by solving certain ordinary differential equations.

Consider for example the *Korteweg–de Vries equation*

$$u_t + uu_x + u_{xxx} = 0, \tag{1.65}$$

that was initially proposed to model the propagation of ocean surface waves that appear due to gravity. Subsequently, it was discovered that (1.65) is also relevant in modeling a large range of phenomena, among which we mention the propagation of heat in solids (the Fermi–Pasta–Ulam model).

The solitons of (1.65) satisfy the third-order ODE

$$\varphi''' + \varphi\varphi' + \mu\varphi' = 0, \tag{1.66}$$

where μ is a real parameter and $\varphi' = \frac{d\varphi}{dx}$.

Note that (1.66) is an equation of type (1.41) and we will use the techniques presented at the end of Sect. 1.2.8 to solve it. We have

$$\varphi'' = \frac{dp}{dx} = \frac{dp}{d\varphi}\frac{d\varphi}{dx} = \frac{dp}{d\varphi}p,$$

and thus

$$\varphi''' = \frac{d}{d\varphi}\left(p\frac{dp}{d\varphi}\right)p.$$

Equation (1.66) reduces to a second-order ODE

$$p\frac{d}{d\varphi}\left(p\frac{dp}{d\varphi}\right) + p\varphi + \mu p = 0,$$

that is,

$$\frac{d}{d\varphi}\left(p\frac{dp}{d\varphi}\right) + \varphi + \mu = 0.$$

Integrating, we deduce that

$$p^2 + \frac{1}{3}\varphi^3 + \mu\varphi^2 + C_1\varphi + C_2 = 0,$$

where C_1 and C_2 are arbitrary constants. We deduce that the solution φ is given by the equation[1]

$$\pm\sqrt{3}\int_{\varphi_0}^{\varphi} \frac{du}{\sqrt{-u^3 - 3\mu u^2 - 3C_1 u - 3C_2}} = t + C_3.*$$

An equation related to (1.65) is

$$u_t + uu_x = \nu u_{xx}. \tag{1.67}$$

It is called the *Burgers equation* and it is used as a mathematical model of turbulence. If we make the change in variables $v_x := u$ and then integrate with respect to x, we obtain the equivalent equation

$$v_t + \frac{1}{2}|v_x|^2 = \nu v_{xx} \tag{1.68}$$

which in its turn has multiple physical interpretations.

The method we employed above can be used to produce explicit solutions to (1.68). More precisely, we seek solutions of the form $v(t,x) = \psi\big(t + y(x)\big)$.

Computing the partial derivatives v_t, v_x and substituting them into (1.68), we deduce that

$$\frac{1 - \nu y''(x)}{y'(x)^2} = \frac{\nu\psi''\big(t + y(x)\big) - \psi'\big(t + y(x)\big)^2/2}{-\psi'\big(t + y(x)\big)}.$$

[1]N.T.: Equality $(*)$ shows that φ is an elliptic function.

Thus, for v to satisfy (1.68) it suffices that y and ψ satisfy the ordinary differential equations

$$C(y')^2 + \nu y'' = 1, \tag{1.69}$$

$$\frac{1}{2}(\psi')^2 - \nu\psi'' = C\psi', \tag{1.70}$$

where C is an arbitrary constant.

To solve (1.69), we set $z := y'$ and we obtain in this fashion the first-order separable equation

$$Cz^2 = 1 - \nu z'$$

which is explicitly solvable. Equation (1.70) can be solved in a similar way and we obtain an explicit solution v of (1.68) and, indirectly, an explicit solution u of (1.67).

1.3.9 Bipartite Biological Systems

We consider here the problem of finding the concentration of a chemical substance (e.g., a medical drug) in a system consisting of two compartments separated by a membrane.

The drug can pass through the membrane in both directions, that is, from compartment **I** to compartment **II** and, reversely, from compartment **II** to compartment **I**, but it can also flow out from compartment **II** to an exterior system. If the compartments have volumes v_1 and v_2, respectively, and $x_1(t)$ and $x_2(t)$ denote the amount of the drug in compartment **I** and **II**, respectively, then the rate $\frac{dx_1}{dt}$ of transfer of the drug from compartment **I** to compartment **II** is proportional to the area A of the membrane and the concentration $\frac{x_1}{v_1}$ of the drug in compartment **I**. Similarly, the rate of transfer of the drug from compartment **II** is proportional to the product $A\frac{x_2}{v_2}$. Therefore, there exist positive constants α, β such that

$$\frac{dx_1}{dt} = -\beta A\frac{x_1}{v_1} + \alpha A\frac{x_2}{v_2}. \tag{1.71}$$

We obtain a similar equation describing the evolution of x_2. More precisely, we have

$$\frac{dx_2}{dt} = \beta A\frac{x_1}{v_1} - \alpha A\frac{x_2}{v_2} - \gamma x_2, \tag{1.72}$$

where γx_2 represents the rate of transfer of the drug from compartment **II** to the exterior system. In particular, γ is also a positive constant.

The differential system (1.71), (1.72) describes the evolution of the amount of the drug in the above bipartite system. This system can be solved by expressing one of the unknowns in terms of the other, thus reducing the system to a second-order ODE. We will discuss more general methods in Chap. 3.

1.3.10 Chemical Reactions

Consider n chemical substances reacting with each other. Let us denote by $x_1(t), \ldots, x_n(t)$ their respective concentrations at time t. The rate of change in the concentration x_i is, in general, a function of the concentrations $x_1, \ldots, x_n$, that is,

$$\frac{dx_i}{dt} = f_i(x_1, \ldots, x_n), \quad i = 1, \ldots, n. \tag{1.73}$$

Let us illustrate this in some special cases.

If the chemical reaction is unimolecular and irreversible of type $A \xrightarrow{k} B$ (the chemical substance A is converted by the reaction into the chemical substance B), then the equation modeling this reaction is

$$-\frac{d}{dt}[A] = k[A] \text{ or } \frac{d}{dt}[B] = k[B], \tag{1.74}$$

where we have denoted by $[A]$ and $[B]$ the concentrations of the substance A and B, respectively.

If the reaction is reversible, $A \underset{k_2}{\overset{k_1}{\rightleftarrows}} B$, then we have the equations

$$\frac{d}{dt}[A] = -k_1[A] + k_2[B], \quad \frac{d}{dt}[B] = k_1[A] - k_2[B],$$

or, if we set $x_1 := [A]$, $x_2 := [B]$, then

$$\frac{dx_1}{dt} = -k_1x_1 + k_2x_2, \quad \frac{dx_2}{dt} = k_1x_1 - k_2x_2.$$

Consider now the case of a bimolecular reaction $A + B \rightarrow P$ in which a moles of the chemical substance A and b moles of the chemical substance B combine to produce the output P.

If we denote by $x(t)$ the number of moles per liter from A and B that enter into the reaction at time t, then according to a well-known chemical law (the law of mass action) the speed of reaction $\frac{dx}{dt}$ is proportional to the product of the concentrations of the chemicals participating in the reaction at time t. In other words,

$$\frac{dx}{dt} = k(x-a)(x-b).$$

Similarly, a chemical reaction involving three chemical substances, $A + B + C \rightarrow P$, is described by the ODE

$$\frac{dx}{dt} = k(x-a)(x-b)(x-c).$$

1.4 Integral Inequalities

This section is devoted to the investigation of the following linear integral inequality

$$x(t) \leq \varphi(t) + \int_a^t \psi(s)x(s)ds, \quad t \in [a, b], \tag{1.75}$$

where the functions x, φ and ψ are continuous on $[a, b]$ and $\psi(t) \geq 0$, $\forall t \in [a, b]$.

Lemma 1.1 (Gronwall's lemma) *If the above conditions are satisfied, then $x(t)$ satisfies the inequality*

$$x(t) \leq \varphi(t) + \int_a^t \varphi(s)\psi(s) \exp\Big(\int_s^t \psi(\tau)d\tau\Big)ds. \tag{1.76}$$

In particular, if $\varphi \equiv C$, (1.76) reduces to

$$x(t) \leq C \exp\Big(\int_a^t \psi(s)ds\Big)ds, \quad \forall\, t \in [a, b]. \tag{1.77}$$

Proof We set

$$y(t) := \int_a^t \psi(s)x(s)ds.$$

Then $y'(t) = \psi(t)x(t)$ and (1.75) can be restated as $x(t) \leq \varphi(t) + y(t)$.
Since $\psi(t) \geq 0$, we have

$$\psi(t)x(t) \leq \psi(t)\varphi(t) + \psi(t)y(t),$$

and we deduce that

$$y'(t) = \psi(t)x(t) \leq \psi(t)\varphi(t) + \psi(t)y(t).$$

We multiply both sides of the above inequality by $\exp\Big(-\int_a^t \psi(s)ds\Big)$ to obtain

$$\frac{d}{dt}\Big(y(t)\exp\Big(-\int_a^t \psi(s)ds\Big)\Big) \leq \psi(t)\varphi(t)\exp\Big(-\int_a^t \psi(s)ds\Big).$$

Integrating, we obtain

$$y(t) \leq \int_a^t \varphi(s)\psi(s) \exp\Big(\int_s^t \psi(\tau)d\tau\Big)ds. \tag{1.78}$$

We reach the desired conclusion by recalling that $x(t) \leq \varphi(t) + y(t)$. □

Corollary 1.1 *Let $x : [a, b] \to \mathbb{R}$ be a continuous nonnegative function satisfying the inequality*

$$x(t) \leq M + \int_a^t \psi(s)x(s)ds, \tag{1.79}$$

where M is a positive constant and $\psi : [a, b] \to \mathbb{R}$ is a continuous nonnegative function. Then

$$x(t) \leq M \exp\Big(\int_a^t \psi(s)ds \Big), \quad \forall t \in [a, b]. \tag{1.80}$$

Remark 1.1 The above inequality is optimal in the following sense: if we have equality in (1.79), then we have equality in (1.80) as well. Note also that we can identify the right-hand side of (1.80) as the unique solution of the linear Cauchy problem

$$x'(t) = \psi(t)x(t), \quad x(a) = M.$$

This Cauchy problem corresponds to the equality case in (1.79).

We will frequently use Gronwall's inequality to produce a priori estimates of solutions of ODEs and systems of ODEs. In the remainder of this section, we will discuss two slight generalizations of this inequality.

Proposition 1.1 (Bihari) *Let $x : [a, b] \to [0, \infty)$ be a continuous function satisfying the inequality*

$$x(t) \leq M + \int_a^t \psi(s)\omega\big(x(s)\big)ds, \quad \forall t \in [a, b], \tag{1.81}$$

where $\omega : [0, \infty) \to (0, \infty)$ is a continuous nondecreasing function. Define $\Phi : [0, \infty) \to \mathbb{R}$ by setting

$$\Phi(u) = \int_{u_0}^u \frac{ds}{\omega(s)} ds, \quad u_0 \geq 0. \tag{1.82}$$

Then

$$x(t) \leq \Phi^{-1}\left(\Phi(M) + \int_a^t \psi(s)ds \right), \quad \forall t \in [a, b]. \tag{1.83}$$

Proof We set

$$y(t) := \int_a^t \omega\big(x(s)\big)\psi(s)ds.$$

Inequality (1.81) implies that $x(t) \leq M + y(t)$, $\forall t \in [a, b]$. Since ω is nondecreasing, we deduce that $y'(t) \leq \omega\big(M + y(t)\big)\psi(t)$. Integrating the last inequality over $[a, t]$, we have

$$\int_M^{y(t)+M} \frac{d\tau}{\omega(\tau)} = \int_{y(a)+M}^{y(t)+M} \frac{d\tau}{\omega(\tau)} = \int_{y(a)}^{y(t)} \frac{ds}{\omega(M+s)} \leq \int_a^t \psi(s)ds,$$

that is,

$$\Phi\big(y(t)+M\big) \leq \Phi(M) + \int_a^t \psi(s)ds.$$

The last inequality is equivalent to inequality (1.83). □

Proposition 1.2 *Let $x : [a, b] \to \mathbb{R}$ be a continuous function that satisfies the inequality*

$$\frac{1}{2}x(t)^2 \leq \frac{1}{2}x_0^2 + \int_a^t \psi(s)|x(s)|ds, \quad \forall t \in [a, b], \tag{1.84}$$

where $\psi : [a, b] \to (0, \infty)$ is a continuous nonnegative function. Then $x(t)$ satisfies the inequality

$$|x(t)| \leq |x_0| + \int_a^t \psi(s)ds, \quad \forall t \in [a, b]. \tag{1.85}$$

Proof For $\varepsilon > 0$ we define

$$y_\varepsilon(t) := \frac{1}{2}(x_0^2 + \varepsilon^2) + \int_a^t \psi(s)|x(s)|ds, \quad \forall t \in [a, b].$$

Using (1.84), we get

$$x(t)^2 \leq 2y_\varepsilon(t), \quad \forall t \in [a, b]. \tag{1.86}$$

Combining this with the equality

$$y_\varepsilon'(t) = \psi(t)|x(t)|$$

and (1.84), we conclude that

$$y_\varepsilon'(t) \leq \sqrt{2y_\varepsilon(t)}\psi(t), \quad \forall t \in [a, b].$$

Integrating from a to t, we obtain

$$\sqrt{2y_\varepsilon(t)} \leq \sqrt{2y_\varepsilon(a)} + \int_a^t \psi(s)ds, \quad \forall t \in [a, b].$$

Using (1.86), we get

$$|x(t)| \leq \sqrt{2y_\varepsilon(a)} + \int_a^t \psi(s)ds \leq |x_0| + \varepsilon + \int_a^t \psi(s)ds, \quad \forall t \in [a, b].$$

Letting $\varepsilon \to 0$ in the above inequality, we obtain (1.85). □

Problems

1.1 A reservoir contains ℓ liters of salt water with the concentration c_0. Salt water is flowing into the reservoir at a rate of ℓ_0-liters per minute and with a concentration α_0. The same amount of salt water is leaving the reservoir every minute. Assuming that the salt in the water is uniformly distributed, find the time evolution of the concentration of salt in the water.
Hint. If we let the state of the system be the concentration $x(t)$ of salt in the water, the data in the problem lead to the ODE

$$\ell x'(t) = \big(\alpha_0 - x(t)\big)\ell_0, \tag{1.87}$$

and the initial condition $x(0) = c_0$. This is a linear and separable ODE that can be solved using the methods outlined in Sect. 1.2.1.

1.2 Prove that any solution of the ODE

$$x' = \sqrt[3]{\frac{x^2+1}{t^4+1}}$$

has two horizontal asymptotes.
Hint. Write the above equation as

$$\int_0^{x(t)} \frac{dy}{\sqrt[3]{x^2+1}} = \int_0^t \sqrt[3]{s^4+1}\,ds$$

and study $\lim\limits_{t\to\pm\infty} x(t)$.

1.3 Find the plane curve with the property that the distance from the origin to any tangent line to the curve is equal to the x-coordinate of the tangency point.

1.4 (Rocket motion) A body of mass m is launched from the surface of the Earth with initial velocity v_0 along the vertical line corresponding to the launching point. Assuming that the air resistance is negligible and taking into account that the gravitational force that acts on the body at altitude x is equal to $\frac{mgR^2}{(x+R)^2}$ (R is the radius of the earth), we deduce from Newton's second law that the altitude $x(t)$ of the body satisfies the ODE

$$x'' = -\frac{gR^2}{(x+R)^2}. \tag{1.88}$$

Solve (1.88) and determine the minimal initial velocity such that the body never returns to Earth.
Hint. Equation (1.88) is of type (1.42) and, using the substitution indicated in Sect. 1.2.8, it can be reduced to a first-order separable ODE.

1.5 Find the solution of the ODE

$$3x^2x' + 16t = 2tx^3$$

that is bounded on the positive semi-axis $[0, \infty)$.
Hint. Via the substitution $x^3 = y$, obtain for y the linear equation $y' - 2ty + 16t = 0$.

1.6 Prove that the ODE

$$x' + \omega x = f(t), \tag{1.89}$$

where ω is a positive constant and $f : \mathbb{R} \to \mathbb{R}$ is continuous and bounded, has a unique solution that is bounded on $\mathbb{R}$. Find this solution x and prove that if f is periodic, then x is also periodic, with the same period.
Hint. Start with the general solution $x(t) = e^{-\omega t}\left(C + \int_{-\infty}^{t} e^{\omega s} f(s)ds\right)$.

1.7 Consider the ODE

$$tx' + ax = f(t),$$

where a is a positive constant and $\lim_{t\to 0} f(t) = \alpha$. Prove that there exists a unique solution of this equation that has finite limit as $t \to 0$ and then find this solution.

1.8 According to Newton's heating and cooling law, the rate of decrease in temperature of a body that is cooling is proportional to the difference between the temperature of the body and the temperature of the ambient surrounding. Find the equation that models the cooling phenomenon.

1.9 Let $f : [0, \infty) \to \mathbb{R}$ be a continuous function such that $\lim_{t\to\infty} f(t) = 0$. Prove that any solution of (1.89) goes to 0 as $t \to \infty$.
Hint. Start with the general solution of the linear ODE (1.89).

1.10 Let $f : [0, \infty) \to \mathbb{R}$ be a continuous function such that $\int_0^\infty |f(t)|dt < \infty$. Prove that the solutions of the ODE

$$x' + \big(\omega + f(t)\big)x = 0, \quad \omega > 0,$$

converge to 0 as $t \to \infty$.

1.11 Solve the (Lotka–Volterra) ODE

$$\frac{dy}{dx} = \frac{y(dx - c)}{x(a - by)}.$$

1.12 Solve the ODE

$$x' = k(a - x)(b - x).$$

(Such an equation models certain chemical reactions.)

1.13 Find the solution of the ODE

$$x' \sin t = 2(x + \cos t)$$

that stays bounded as $t \to \infty$.
Hint. Solve it as a linear ODE.

1.14 Prove that, by using the substitution $y = \frac{x'}{x}$, we can reduce the second-order ODE $x'' = a(t)x$ to the Riccati-type equation

$$y' = -y^2 + a(t). \tag{1.90}$$

1.15 Prove that, if $x_1(t), x_2(t), x_3(t), x_4(t)$ are solutions of a Riccati-type ODE, then the cross-ratio

$$\frac{x_3(t) - x_1(t)}{x_3(t) - x_2(t)} : \frac{x_4(t) - x_1(t)}{x_4(t) - x_2(t)}$$

is independent of t.

1.16 Find the plane curves such that the area of the triangle formed by any tangent with the coordinate axes is a given constant a^2.

1.17 Consider the family of curves in the (t, x)-plane described by

$$F(t, x, \lambda) = 0, \quad \lambda \in \mathbb{R}. \tag{1.91}$$

(a) Find the curves that are orthogonal to all the curves in this family.
(b) Find the curves that are orthogonal to all those in the family $x = \lambda e^t$.

Hint. Since the tangent line to a curve is parallel to the vector $\left(1, -\frac{F_x}{F_t}\right)$, the orthogonal curves are solutions to the differential equation

$$x' + \frac{F_x}{F_t} = 0, \tag{1.92}$$

where λ has been replaced by its value $\lambda = \lambda(t, x)$ determined from (1.91).

1.18 Find the solitons of the Klein–Gordon equation

$$u_{tt} - u_{xx} + u + u^3 = 0. \tag{1.93}$$

Hint. For $u(x, t) = \varphi(x + \mu t)$ we get for φ the ODE $(\mu^2 - 1)\varphi'' + \varphi + \varphi^3 = 0$.

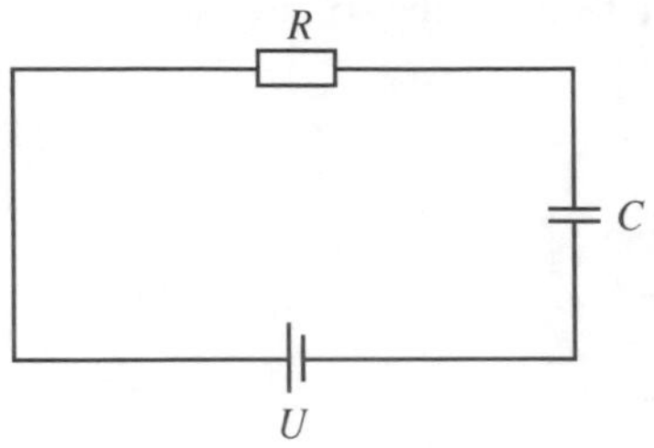

Fig. 1.3 An RC circuit

1.19 Find the differential equation that models the behavior of an RC electric circuit as in Fig. 1.3.
Hint. If we denote by Q the electric charge of the capacitor, then we have $C^{-1}Q + RI = U$, where I denotes the electric current. Thus Q satisfies the ODE

$$R\frac{dQ}{dt} + C^{-1}Q = U. \tag{1.94}$$

1.20 Find the system of ODEs that models the behavior of the electrical circuit in Fig. 1.4.
Hint. Denote by Q_i the electrical charge of the capacitor C_i, $i = 1, 2$, and by I_i the corresponding electrical currents. Kirchoff's laws yield the equations

$$\begin{aligned} C_2^{-1}Q_2 + RI_1 = U_2, \quad -C_2^{-1}Q_2 + R_1I_2 + C_1^{-1}Q_1 = 0,\\ \frac{dQ_1}{dt} = I_2, \quad \frac{dQ_2}{dt} = I_1 - I_2. \end{aligned} \tag{1.95}$$

Using as a state of the system the pair $x_1 = Q_1$ and $x_2 = Q_2$, we obtain a system of first-order ODEs in (x_1, x_2).

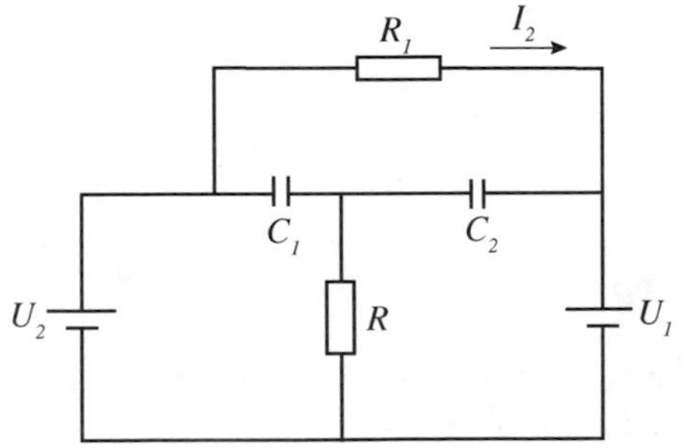

Fig. 1.4 A more complex RC circuit

Chapter 2
Existence and Uniqueness for the Cauchy Problem

In this chapter, we will present some results concerning the existence, uniqueness and dependence on data of the solutions to the Cauchy problem for ODEs and systems of ODEs. From a mathematical point of view, this is a fundamental issue in the theory of differential equations. If we view a differential equation as a mathematical model of a physical theory, the existence of solutions to the Cauchy problem represents one of the first means of testing the validity of the model and, ultimately, of the physical theory. An existence result highlights the states and the minimal physical parameters that determine the evolution of a process and, often having a constructive character, it leads to numerical procedures for approximating the solutions. Basic references for this chapter are [1, 6, 9, 11, 18].

2.1 Existence and Uniqueness for First-Order ODEs

We begin by investigating the existence and uniqueness of solutions to a Cauchy problem in a special case, namely that of the scalar ODE (1.2) defined in a rectangle centered at $(t_0, x_0) \in \mathbb{R}^2$. In other words, we consider the Cauchy problem

$$x' = f(t, x), \quad x(t_0) = x_0, \tag{2.1}$$

where f is a real-valued function defined in the domain

$$\Delta := \Big\{ (t, x) \in \mathbb{R}^2; \quad |t - t_0| \leq a, \quad |x - x_0| \leq b \Big\}. \tag{2.2}$$

The central existence result for problem (2.1) is stated in our next theorem.

V. Barbu, *Differential Equations*, Springer Undergraduate Mathematics Series,
DOI 10.1007/978-3-319-45261-6_2

Theorem 2.1 *Assume that the following hold:*

(i) *The function f is continuous on Δ.*
(ii) *The function f satisfies the Lipschitz condition in the variable x, that is, there exists an $L > 0$ such that*

$$\big|\, f(t,x) - f(t,y) \,\big| \le L|x-y|,\ \ \forall (t,x),\ (t,y) \in \Delta. \tag{2.3}$$

Then there exists a unique solution $x = x(t)$ to the Cauchy problem (2.1) *defined on the interval $|t - t_0| \le \delta$, where*

$$\delta := \min\left(a, \frac{b}{M}\right),\ \ M := \sup_{(t,x)\in\Delta} |f(t,x)|. \tag{2.4}$$

Proof We begin by observing that problem (2.1) is equivalent to the integral equation

$$x(t) = x_0 + \int_{t_0}^{t} f\big(s, x(s)\big)ds. \tag{2.5}$$

Indeed, if the continuous function $x(t)$ satisfies (2.5) on an interval I, then it is clearly a C^1-function and satisfies the initial condition $x(t_0) = x_0$. The equality $x'(t) = f\big(t, x(t)\big)$ is then an immediate consequence of the Fundamental Theorem of Calculus. Conversely, any solution of (2.1) is also a solution of (2.5). Hence, to prove the theorem it suffices to show that (2.5) has a unique continuous solution on the interval $I := [t_0 - \delta, t_0 + \delta]$.

We will rely on the method of successive approximations used by many mathematicians, starting with Newton, to solve algebraic and transcendental equations. For the problem at hand, this method was successfully pioneered by *E. Picard* (1856–1941).

Consider the sequence of functions $x_n : I \to \mathbb{R}$, $n = 0, 1, \ldots,$ defined iteratively as follows

$$\begin{aligned} x_0(t) &= x_0,\ \ \forall t \in I, \\ x_{n+1}(t) &= x_0 + \int_{t_0}^{t} f(s, x_n(s))ds,\ \ \forall t \in I,\ \ \forall n = 0, 1, \ldots. \end{aligned} \tag{2.6}$$

It is easy to see that the functions x_n are continuous and, moreover,

$$\big|\, x_n(t) - x_0 \,\big| \le M\delta \le b,\ \ \forall t \in I,\ \ n = 1, 2, \ldots. \tag{2.7}$$

This proves that the sequence $\{x_n\}_{n\ge 0}$ is well defined. We will prove that this sequence converges uniformly to a solution of (2.5). Using (2.6) and the Lipschitz condition (2.3), we deduce that

$$\begin{aligned}\big|\,x_n(t) - x_{n-1}(t)\,\big| &\leq \int_{t_0}^{t} \big|\,f(s, x_{n-1}(s)) - f(s, x_{n-2}(s)\,)\big|ds \\ &\leq L\left|\int_{t_0}^{t} \big|\,x_{n-1}(s) - x_{n-2}(s)\,\big|ds\right|.\end{aligned} \tag{2.8}$$

Iterating (2.8) and using (2.7), we have

$$\big|\,x_n(t) - x_{n-1}(t)\,\big| \leq \frac{ML^{n-1}}{n!}|t - t_0|^n \leq \frac{ML^{n-1}\delta^n}{n!}, \quad \forall n, \quad \forall t \in I. \tag{2.9}$$

Observe that the sequence $\{x_n\}_{n\geq 0}$ is uniformly convergent on I if and only if the telescopic series

$$\sum_{n\geq 1}\big(\,x_n(t) - x_{n-1}(t)\,\big)$$

is uniformly convergent on this interval. The uniform convergence of this series follows from (2.9) by invoking Weierstrass' M-test: the above series is majorized by the convergent numerical series

$$\sum_{n\geq 1}\frac{ML^{n-1}\delta^n}{n!}.$$

Hence the limit

$$x(t) = \lim_{n\to\infty} x_n(t)$$

exists uniformly on the interval I. The function $x(t)$ is continuous, and from the uniform continuity of the function $f(t,x)$ we deduce that

$$f\big(t, x(t)\big) = \lim_{n\to\infty} f\big(t, x_n(t)\big),$$

uniformly in $t \in I$. We can pass to the limit in the integral that appears in (2.6) and we deduce that

$$x(t) = x_0 + \int_{t_0}^{t} f\big(s, x(s)\big)ds, \quad \forall t \in I. \tag{2.10}$$

In other words, $x(t)$ is a solution of (2.5).

To prove the uniqueness, we argue by contradiction and assume that $x(t), y(t)$ are two solutions of (2.5) on I. Thus

$$\big|x(t) - y(t)\big| = \left|\int_{t_0}^{t} f\big(s, x(s)\big) - f\big(s, y(s)\big)ds\right| \leq L\left|\int_{t_0}^{t} |x(s) - y(s)|ds\right|, \quad \forall t \in I.$$

Using Gronwall's Lemma 1.1 with $\varphi \equiv 0$ and $\psi \equiv L$, we deduce $x(t) = y(t)$, $\forall t \in I$. □

Remark 2.1 In particular, the Lipschitz condition (2.3) is satisfied if the function f has a partial derivative $\frac{\partial f}{\partial x}$ that is continuous on the rectangle Δ, or more generally, that is bounded on this rectangle.

Remark 2.2 We note that Theorem 2.1 is a local existence and uniqueness result for the Cauchy problem (2.1), that is the existence and uniqueness was proved on an interval $[t_0 - \delta, t_0 + \delta]$ which is smaller than the interval $[t_0 - a, t_0 + a]$ of the definition for the functions $t \to f(t, x)$.

2.2 Existence and Uniqueness for Systems of First-Order ODEs

Consider the differential system

$$x_i' = f_i(t, x_1, \dots, x_n), \quad i = 1, \dots, n, \tag{2.11}$$

together with the initial conditions

$$x_i(t_0) = x_i^0, \quad i = 1, \dots, n, \tag{2.12}$$

where the functions f_i are defined on a parallelepiped

$$\Delta := \{(t, x_1, \dots, x_n) \in \mathbb{R}^{n+1};\ |t - t_0| \le a,\ |x_i - x_i^0| \le b,\ i = 1, \dots, n\}. \tag{2.13}$$

Theorem 2.1 generalizes to differential systems of type (2.11).

Theorem 2.2 *Assume that the following hold:*

(i) *The functions f_i are continuous on Δ for any $i = 1, \dots, n$.*
(ii) *The functions f_i are Lipschitz in $\boldsymbol{x} = (x_1, \dots, x_n)$ on Δ, that is, there exists an $L > 0$ such that*

$$\big| f_i(t, x_1, \dots, x_n) - f_i(t, y_1, \dots, y_n) \big| \le L \max_{1 \le k \le n} |x_k - y_k|, \tag{2.14}$$

for any $i = 1, \dots, n$ and any $(t, x_1, \dots, x_n), (t, y_1, \dots, y_n) \in \Delta$.

Then there exists a unique solution $x_i = \varphi_i(t)$, $i = 1, \dots, n$, of the Cauchy problem (2.11) and (2.12) *defined on the interval*

$$I := [t_0 - \delta, t_0 + \delta], \quad \delta := \min\left(a, \frac{b}{M}\right), \tag{2.15}$$

where $M := \max\big\{ |f_i(t, \boldsymbol{x})|;\ (t, \boldsymbol{x}) \in \Delta,\ i = 1, \dots, n \big\}$.

Proof The proof of Theorem 2.2 is based on an argument very similar to the one used in the proof of Theorem 2.1. For this reason, we will only highlight the main steps.

We observe that the Cauchy problem (2.11) and (2.12) is equivalent to the system of integral equations

$$x_i(t) = x_i^0 + \int_a^t f_i\big(s, x_1(s), \dots, x_n(s)\big)ds, \quad i = 1, \dots, n. \tag{2.16}$$

To construct a solution to this system, we again use successive approximations

$$\begin{aligned} &x_i^k(t) = x_i^0 + \int_a^t f_i\big(s, x_1^{k-1}(s), \dots, x_n^{k-1}(s)\big)ds, \ i = 1, \dots, n, \ k \geq 1, \\ &x_i^0(s) \equiv x_i^0, \ \ i = 1, \dots, n. \end{aligned} \tag{2.17}$$

Arguing as in the proof of Theorem 2.1, we deduce that the functions $t \mapsto x_i^k(t)$ are well defined and continuous on the interval I. An elementary argument based on the Lipschitz condition yields the following counterpart of (2.9)

$$\max_{1 \leq i \leq n} |x_i^k(t)| \leq \frac{ML^{k-1}\delta^k}{k!}, \quad \forall k \geq 1, \ \ t \in I.$$

Invoking as before the Weierstrass M-test, we deduce that the limits

$$\varphi_i(t) = \lim_{k \to \infty} x_i^k(t), \quad 1 \leq i \leq n,$$

exist and are uniform on I. Letting $k \to \infty$ in (2.17), we deduce that $(\varphi_1, \dots, \varphi_n)$ is a solution of system (2.16), and thus also a solution of the Cauchy problem (2.11) and (2.12).

The uniqueness follows from Gronwall's Lemma (Lemma 1.1) via an argument similar to the one in the proof of Theorem 2.1. □

Both the statement and the proof of the existence and uniqueness theorem for systems do not seem to display meaningful differences when compared to the scalar case. Once we adopt the vector notation, we will see that there are not even formal differences between these two cases.

Consider the vector space $\mathbb{R}^n$ of vectors $\boldsymbol{x} = (x_1, \dots, x_n)$ equipped with the norm (see Appendix A)

$$\|\boldsymbol{x}\| := \max_{1 \leq i \leq n} |x_i|, \quad \boldsymbol{x} = (x_1, \dots, x_n). \tag{2.18}$$

On the space $\mathbb{R}^n$ equipped with the above (or any other) norm, we can develop a differential and integral calculus similar to the familiar one involving scalar functions.

Given an interval I, we define a vector-valued function $\boldsymbol{x} : I \to \mathbb{R}$ of the form

$$\boldsymbol{x}(t) = \big(x_1(t), \dots, x_n(t)\big),$$

where $x_i(t)$ are scalar functions defined on I. The function $\boldsymbol{x} : I \to \mathbb{R}^n$ is called *continuous* if all its components $\{x_i(t);\ i = 1, \dots, n\}$ are continuous. The function $\boldsymbol{x}$ is called *differentiable* at t_0 if all its components x_i have this property. The derivative of $\boldsymbol{x}(t)$ at the point t, denoted by $\boldsymbol{x}'(t)$, is the vector

$$\boldsymbol{x}'(t) := \big(x_1'(t), \dots, x_n'(t) \big).$$

We can define the integral of the vector function in a similar fashion. More precisely,

$$\int_a^b \boldsymbol{x}(t)dt := \left(\int_a^b x_1(t)dt, \dots, \int_a^b x_n(t)dt \right) \in \mathbb{R}^n.$$

The sequence $\{\boldsymbol{x}^\nu\}$ of vector-valued functions

$$\boldsymbol{x}^\nu : I \to \mathbb{R}^n, \quad \nu = 0, 1, 2, \dots,$$

is said to *converge uniformly* (respectively *pointwisely*) to $\boldsymbol{x} : I \to \mathbb{R}^n$ as $\nu \to \infty$ if each component sequence has these properties. The space of continuous functions $x : I \to \mathbb{R}^n$ is denoted by $C(I; \mathbb{R}^n)$.

All the above notions have an equivalent formulation involving the norm $\| - \|$ of the space $\mathbb{R}^n$; see Appendix A. For example, the continuity of $\boldsymbol{x} : I \to \mathbb{R}^n$ at $t_0 \in I$ is equivalent to

$$\lim_{t \to t_0} \|\boldsymbol{x}(t) - \boldsymbol{x}(t_0)\| = 0.$$

The derivative, integral and the concept of convergence can be defined along similar lines.

Returning to the differential system (2.11), observe that, if we denote by $\boldsymbol{x}(t)$ the vector-valued function

$$\boldsymbol{x}(t) = (\, x_1(t), \dots, x_n(t) \,)$$

and by $\boldsymbol{f} : \Delta \to \mathbb{R}^n$ the function

$$\boldsymbol{f}(t, \boldsymbol{x}) = \big(f_1(t, \boldsymbol{x}), \dots, f_n(t, \boldsymbol{x}) \big),$$

then we can rewrite (2.11) as

$$\boldsymbol{x}' = \boldsymbol{f}(t, \boldsymbol{x}), \tag{2.19}$$

while the initial condition (2.12) becomes

$$\boldsymbol{x}(t_0) = \boldsymbol{x}^0 := (x_1^0, \dots, x_n^0). \tag{2.20}$$

In vector notation, Theorem 2.2 can be rephrased as follows.

Theorem 2.3 *Assume that the following hold:*

(i) *The function* $\boldsymbol{f} : \Delta \to \mathbb{R}^n$ *is continuous.*
(ii) *The function* $\boldsymbol{f}$ *is Lipschitz in the variable* $\boldsymbol{x}$ *on* Δ.

Then there exists a unique solution $\boldsymbol{x} = \varphi(t)$ *of the system* (2.19) *satisfying the initial condition* (2.20) *and defined on the interval*

$$I := [t_0 - \delta, t_0 + \delta], \quad \delta := \min\left(a, \frac{b}{M}\right), \quad M := \sup_{(t,\boldsymbol{x})\in\Delta} \|\boldsymbol{f}(t,\boldsymbol{x})\|.$$

In this formulation, Theorem 2.3 can be proved by following word for word the proof of Theorem 2.1, with one obvious exception: where appropriate, we need to replace the absolute value $| - |$ with the norm $\| - \|$. In the sequel, we will systematically use the vector notation when working with systems of differential equations.

2.3 Existence and Uniqueness for Higher Order ODEs

Consider the differential equation of order n,

$$x^{(n)} = g(t, x, x', \dots, x^{(n-1)}), \tag{2.21}$$

together with the Cauchy condition (see Sect. 1.1)

$$x(t_0) = x_0^0, \quad x'(t_0) = x_1^0, \dots, \quad x^{(n-1)}(t_0) = x_{n-1}^0, \tag{2.22}$$

where $(t_0, x_0^0, x_1^0, \dots, x_{n-1}^0) \in \mathbb{R}^{n+1}$ is fixed and the function g satisfies the following conditions.

(I) The function g is defined and continuous on the set

$$\Delta = \left\{ (t, x_1, \dots, x_n) \in \mathbb{R}^{n+1}; \quad |t - t_0| \le a, \quad |x_i - x_{i-1}^0| \le b, \quad \forall i = 1, \dots, n \right\}.$$

(II) There exists an $L > 0$ such that

$$|g(t, \boldsymbol{x}) - g(t, \boldsymbol{y})| \le L\|\boldsymbol{x} - \boldsymbol{y}\|, \quad \forall (t, \boldsymbol{x}), \ (t, \boldsymbol{y}) \in \Delta. \tag{2.23}$$

Theorem 2.4 *Assume that conditions* (I) *and* (II) *above hold. Then the Cauchy problem* (2.21) *and* (2.22) *admits a unique solution on the interval*

$$I := [t_0 - \delta, t_0 + \delta], \quad \delta := \min\left(a, \frac{b}{M}\right),$$

where

$$M := \sup_{(t,x)\in\Delta} \max\big\{\, |g(t,\boldsymbol{x})|, |x_2|, \dots, |x_n| \,\big\}.$$

Proof As explained before (see (1.10) and (1.11)), using the substitutions

$$x_1 := x, \quad x_2 = x', \dots, x_n := x^{(n-1)},$$

the differential equation (2.21) reduces to the system of ODEs

$$\begin{aligned} x_1' &= x_2 \\ x_2' &= x_3 \\ \vdots\; &\;\vdots\; \vdots \\ x_n' &= g(t, x_1, \dots, x_n), \end{aligned} \tag{2.24}$$

while the Cauchy condition becomes

$$x_i(t_0) = x_{i-1}^0, \quad \forall i = 1, \dots, n. \tag{2.25}$$

In view of (I) and (II), Theorem 2.4 becomes a special case of Theorem 2.2. □

2.4 Peano's Existence Theorem

We will prove an existence result for the Cauchy problem due to *G. Peano* (1858–1932). Roughly speaking, it states that the continuity of f alone suffices to guarantee that the Cauchy problem (2.11) and (2.12) has a solution in a neighborhood of the initial point. Beyond its theoretical significance, this result will offer us the opportunity to discuss another important technique for investigating and approximating the solutions of an ODE. We are talking about the polygonal method, due essentially to *L. Euler* (1707–1783).

Theorem 2.5 *Let $f : \Delta \to \mathbb{R}^n$ be a continuous function defined on*

$$\Delta := \big\{\, (t, \boldsymbol{x}) \in \mathbb{R}^{n+1};\ \ |t - t_0| \le a,\ \ \|\boldsymbol{x} - \boldsymbol{x}_0\| \le b \,\big\}.$$

Then the Cauchy problem (2.11) *and* (2.12) *admits at least one solution on the interval*

$$I := [t_0 - \delta, t_0 + \delta],\quad \delta := \min\left(a, \frac{b}{M}\right),\quad M := \sup_{(t,\boldsymbol{x})\in\Delta} \|\boldsymbol{f}(t, \boldsymbol{x})\|.$$

Proof We will prove the existence on the interval $[t_0, t_0 + \delta]$. The existence on $[t_0 - \delta, t_0]$ follows by a similar argument.

Fix $\varepsilon > 0$. Since f is uniformly continuous on Δ, there exists an $\eta(\varepsilon) > 0$ such that

$$\|f(t, \boldsymbol{x}) - f(s, \boldsymbol{y})\| \leq \varepsilon,$$

for any $(t, \boldsymbol{x}), (s, \boldsymbol{y}) \in \Delta$ such that

$$|t - s| \leq \eta(\varepsilon), \quad \|\boldsymbol{x} - \boldsymbol{y}\| \leq \eta(\varepsilon).$$

Consider the uniform subdivision $t_0 < t_1 < \cdots < t_{N(\varepsilon)} = t_0 + \delta$, where $t_j = t_0 + jh_\varepsilon$, for $j = 0, \ldots, N(\varepsilon)$, and $N(\varepsilon)$ is chosen large enough so that

$$h_\varepsilon = \frac{\delta}{N(\varepsilon)} \leq \min\left(\eta(\varepsilon), \frac{\eta(\varepsilon)}{M}\right). \tag{2.26}$$

We consider the polygonal line, that is, the piecewise linear function $\varphi_\varepsilon : [t_0, t_0 + \delta] \to \mathbb{R}^{n+1}$ defined by

$$\begin{aligned} &\varphi_\varepsilon(t) = \varphi_\varepsilon(t_j) + (t - t_j)\boldsymbol{f}\big(t, \varphi_\varepsilon(t_j)\big), \ t_j < t \leq t_{j+1} \\ &\varphi_\varepsilon(t_0) = \boldsymbol{x}_0. \end{aligned} \tag{2.27}$$

Notice that if $t \in [t_0, t_0 + \delta]$, then

$$\|\varphi_\varepsilon(t) - \boldsymbol{x}_0\| \leq M\delta \leq b.$$

Thus $(t, \varphi_\varepsilon(t)) \in \Delta$, $\forall t \in [t_0, t_0 + \delta]$, so that equalities (2.27) are consistent. Equalities (2.27) also imply the estimates

$$\|\varphi_\varepsilon(t) - \varphi_\varepsilon(s)\| \leq M|t - s|, \quad \forall t, x \in [t_0, t_0 + \delta]. \tag{2.28}$$

In particular, inequality (2.28) shows that the family of functions $(\varphi_\varepsilon)_{\varepsilon>0}$ is uniformly bounded and equicontinuous on the interval $[t_0, t_0 + \delta]$. Arzelà's theorem (see Appendix A.3) shows that there exist a continuous function $\varphi : [t_0, t_0 + \delta] \to \mathbb{R}^n$ and a subsequence $(\varphi_{\varepsilon_\nu})$, $\varepsilon_\nu \searrow 0$, such that

$$\lim_{\nu\to\infty} \varphi_{\varepsilon_\nu}(t) = \varphi(t) \ \text{ uniformly on } [t_0, t_0 + \delta]. \tag{2.29}$$

We will prove that $\varphi(t)$ is a solution of the Cauchy problem (2.11) and (2.12).

With this goal in mind, we consider the sequence of functions

$$g_{\varepsilon_\nu}(t) := \begin{cases} \varphi'_{\varepsilon_\nu}(t) - \boldsymbol{f}(t, \varphi_{\varepsilon_\nu}(t)), & t \neq t_j^\nu \\ 0, & t = t_j^\nu, \quad j = 0, 1, \ldots, N(\varepsilon_\nu), \end{cases} \tag{2.30}$$

where t_j^ν, $j = 0, 1, \ldots, N(\varepsilon_\nu)$, are the nodes of the subdivision corresponding to ε_ν. Equality (2.27) implies that

$$\varphi'_{\varepsilon_\nu}(t) = f\big(t,\ \varphi_{\varepsilon_\nu}(t_j^\nu)\big),\ \ \forall t \in]t_j^\nu, t_{j+1}^\nu[,$$

and thus, invoking (2.26), we deduce that

$$\|g_{\varepsilon_\nu}(t)\| \leq \varepsilon_\nu,\ \ \forall t \in [t_0, t_0 + \delta]. \tag{2.31}$$

On the other hand, the function g_{ε_ν}, though discontinuous, is Riemann integrable on $[t_0, t_0 + \delta]$ since its set of discontinuity points, $\{t_j^\nu\}_{0 \leq j \leq N(\varepsilon_\nu)}$, is finite. Integrating both sides of (2.30), we get

$$\varphi_{\varepsilon_\nu}(t) = x_0 + \int_{t_0}^t f\big(s, \varphi_{\varepsilon_\nu}(s)\big)ds + \int_{t_0}^t g_{\varepsilon_\nu}(s)ds,\ \ \forall t \in [t_0, t_0 + \delta]. \tag{2.32}$$

Since f is continuous on Δ and $\varphi_{\varepsilon_\nu}$ converge uniformly on $[t_0, t_0 + \delta]$, we have

$$\lim_{\nu \to \infty} f\big(s, \varphi_{\varepsilon_\nu}(s)\big) = f\big(s, \varphi(s)\big)\ \text{ uniformly in } s \in [t_0, t_0 + \delta].$$

Invoking (2.31), we can pass to the limit in (2.32) and we obtain the equality

$$\varphi(t) = x_0 + \int_{t_0}^t f\big(s, \varphi(s)\big)ds.$$

In other words, the function $\varphi(t)$ is a solution of the Cauchy problem (2.11) and (2.12). This completes the proof of Theorem 2.5. □

An alternative proof. Consider a sequence $f_\varepsilon : \Delta \to \mathbb{R}^n$ of continuously differentiable functions such that

$$\|f_\varepsilon(t, x) - f_\varepsilon(t, y)\| \leq L_\varepsilon \|x - y\|,\ \ \forall\ (t, x), (t, y) \in \Delta, \tag{2.33}$$

$$\lim_{\varepsilon \to 0} \|f_\varepsilon(t, x) - f(t, x)\| = 0,\ \ \text{uniformly on } \Delta. \tag{2.34}$$

An example of such an approximation f_ε of f is

$$f_\varepsilon(t, x) = \frac{1}{\varepsilon^n} \int_{[y;\ \|y - x_0\| \leq b]} f(t, y)\rho\left(\frac{x - y}{\varepsilon}\right) dy,\ \ \forall\ (t, x) \in \Delta,$$

where $\rho : \mathbb{R}^n \to \mathbb{R}$ is a differentiable function such that $\int_{\mathbb{R}^n} \rho(x)dx = 1,\ \rho(x) = 0$ for $\|x\| \geq 1$.) Then, by Theorem 2.3, the Cauchy problem

$$\begin{aligned} &\frac{dx}{dt}(t) = f_\varepsilon(t, x(t)),\ t \in [t_0 = \delta, t_0 + \delta), \\ &x(t_0) = x_0, \end{aligned} \tag{2.35}$$

has a unique solution $\boldsymbol{x}_\varepsilon$ on the interval $[t_0-\delta, t_0+\delta)$. By (2.34) and (2.35), it follows that

$$\left\| \frac{d}{dt}\, \boldsymbol{x}_\varepsilon(t) \right\| \le C, \quad \forall\, t \in [t_0-\delta, t_0+\delta], \quad \forall\, \varepsilon > 0,$$

where C is independent of ε. This implies that the family of functions $\{\boldsymbol{x}_\varepsilon\}$ is uniformly bounded and equicontinuous on $[t_0-\delta, t_0+\delta]$ and so, by Arzelà's theorem, there is a subsequence $\{\boldsymbol{x}_{\varepsilon_n}\}$ which is uniformly convergent for $\{\varepsilon_n\} \to 0$ to a continuous function $\boldsymbol{x} : [t_0-\delta, t_0+\delta]$. Then, by (2.35), it follows that

$$\boldsymbol{x}(t) = \boldsymbol{x}_0 + \int_{t_0}^{t} \boldsymbol{f}(s, \boldsymbol{x}(s))ds, \quad \forall\, t \in [t_0-\delta, t_0+\delta]$$

and so $\boldsymbol{x}$ is a solution to the Cauchy problem (2.11) and (2.12).

Remark 2.3 (*Nonuniqueness in the Cauchy problem*) We cannot deduce the uniqueness of the solution from the above proof since the family $(\varphi_\varepsilon)_{\varepsilon>0}$ may contain several uniform convergent subsequences, each with its own limit. In general, assuming only the continuity of $\boldsymbol{f}$, we cannot expect uniqueness in the Cauchy problem. One example of nonuniqueness is offered by the Cauchy problem

$$x' = x^{\frac{1}{3}}, \quad x(0) = 0. \tag{2.36}$$

This equation has an obvious solution $x(t) = 0, \forall t$. On the other hand, as easily seen, the function

$$\varphi(t) = \begin{cases} \left(\dfrac{2t}{3}\right)^{\frac{3}{2}}, & t \ge 0, \\ 0, & t < 0, \end{cases}$$

is also a solution of (2.36).

Remark 2.4 (*Numerical approximations*) If, in Theorem 2.5, we assume that $\boldsymbol{f} = \boldsymbol{f}(t, \boldsymbol{x})$ is Lipschitz in $\boldsymbol{x}$, then according to Theorem 2.3 the Cauchy problem (2.19)–(2.20) has a *unique* solution. Thus, necessarily,

$$\lim_{\varepsilon \searrow 0} \varphi_\varepsilon(t) = \varphi(t) \;\; \text{uniformly on } [t_0-\delta, t_0+\delta], \tag{2.37}$$

because any sequence of the family $(\varphi_\varepsilon)_{\varepsilon>0}$ contains a subsequence that converges uniformly to $\varphi(t)$. Thus, the above procedure leads to a numerical approximation scheme for the solution of the Cauchy problem (2.19)–(2.20), or equivalently (2.11) and (2.12). If h is fixed, $h = \frac{\delta}{N}$, and

$$t_j := t_0 + jh, \quad j = 0, 1, \ldots, N,$$

then we compute the approximations of the values of $\varphi(t)$ at the nodes t_j using (2.27), that is,

$$\varphi_{j+1} = \varphi_j + hf\big(t_j, \varphi_j\big), \quad j = 0, 1, \dots, N-1. \tag{2.38}$$

The iterative formulae (2.38) are known in numerical analysis as the *Euler scheme* and they form the basis of an important class of numerical methods for solving the Cauchy problem. Equalities (2.38) are also known as *difference equations*.

2.5 Global Existence and Uniqueness

We consider the system of differential equations described in vector notation by

$$\boldsymbol{x}' = \boldsymbol{f}(t, \boldsymbol{x}), \tag{2.39}$$

where the function $\boldsymbol{f} : \Omega \to \mathbb{R}^n$ is continuous on the open subset $\Omega \subset \mathbb{R}^{n+1}$. Additionally, we will assume that $\boldsymbol{f}$ is *locally Lipschitz* in $\boldsymbol{x}$ on Ω, that is, for any compact set $K \subset \Omega$, there exists an $L_K > 0$ such that

$$\| \boldsymbol{f}(t, \boldsymbol{x}) - \boldsymbol{f}(t, \boldsymbol{y}) \| \leq L_K \|\boldsymbol{x} - \boldsymbol{y}\|, \quad \forall (t, \boldsymbol{x}), (t, \boldsymbol{y}) \in K. \tag{2.40}$$

If $A, B \subset \mathbb{R}^m$, then the distance between them is defined by

$$\operatorname{dist}(A, B) = \inf\big\{ \|\boldsymbol{a} - \boldsymbol{b}\|; \;\; \boldsymbol{a} \in A, \;\; \boldsymbol{b} \in B \big\}.$$

It is useful to remark that, if K is a compact subset of Ω, then the distance $\operatorname{dist}(K, \partial\Omega)$ from K to the boundary $\partial\Omega$ of Ω is strictly positive. Indeed, suppose that $(\boldsymbol{x}_\nu)$ is a sequence in K and $(\boldsymbol{y}_\nu)$ is a sequence in $\partial\Omega$ such that

$$\lim_{\nu\to\infty} \|\boldsymbol{x}_\nu - \boldsymbol{y}_\nu\| = \operatorname{dist}(K, \partial\Omega). \tag{2.41}$$

Since K is compact, the sequence $(\boldsymbol{x}_\nu)$ is bounded. Using (2.41), we deduce that the sequence $(\boldsymbol{y}_\nu)$ is also bounded. The Bolzano–Weierstrass theorem now implies that there exist subsequences $(\boldsymbol{x}_{\nu_k})$ and $(\boldsymbol{y}_{\nu_k})$ converging to $\boldsymbol{x}_0$ and respectively $\boldsymbol{y}_0$. Since both K and $\partial\Omega$ are closed, we deduce that $\boldsymbol{x}_0 \in K$, $\boldsymbol{y}_0 \in \partial\Omega$, and

$$\|\boldsymbol{x}_0 - \boldsymbol{y}_0\| = \lim_{k\to\infty} \|\boldsymbol{x}_{\nu_k} - \boldsymbol{y}_{\nu_k}\| = \operatorname{dist}(K, \partial\Omega).$$

Since $K \cap \partial\Omega = \emptyset$, we conclude that $\operatorname{dist}(K, \partial\Omega) > 0$.

Returning to the differential system (2.39), consider $(t_0, \boldsymbol{x}_0) \in \Omega$ and a parallelepiped $\Delta \subset \Omega$ of the form

$$\Delta = \Delta_{a,b} := \{ (t, \boldsymbol{x}) \in \mathbb{R}^{n+1};\ \ |t - t_0| \leq a,\ \ \|\boldsymbol{x} - \boldsymbol{x}_0\| \leq b \}.$$

(Since Ω is open, $\Delta_{a,b} \subset \Omega$ if a and b are sufficiently small.)

Applying Theorem 2.3 to system (2.39) restricted to Δ, we deduce the existence and uniqueness of a solution $\boldsymbol{x} = \varphi(t)$ satisfying the initial condition $\varphi(t_0) = \boldsymbol{x}_0$ and defined on an interval $[t_0 - \delta, t_0 + \delta]$, where

$$\delta = \min\left(a, \frac{b}{M}\right),\quad M = \sup_{(t,\boldsymbol{x})\in\Delta} \|\boldsymbol{f}(t, \boldsymbol{x})\|.$$

In other words, we have the following local existence result.

Theorem 2.6 *Let $\Omega \subset \mathbb{R}^{n+1}$ be an open set and assume that the function $\boldsymbol{f} = \boldsymbol{f}(t, \boldsymbol{x}) : \Omega \to \mathbb{R}^n$ is continuous and locally Lipschitz as a function of $\boldsymbol{x}$. Then for any $(t_0, \boldsymbol{x}_0) \in \Omega$ there exists a unique solution $\boldsymbol{x}(t) = \boldsymbol{x}(t; t_0, \boldsymbol{x}_0)$ of (2.39) defined on a neighborhood of t_0 and satisfying the initial condition*

$$\boldsymbol{x}(t; t_0, \boldsymbol{x}_0)\Big|_{t=t_0} = \boldsymbol{x}_0.$$

We must emphasize the local character of the above result. As mentioned earlier in Remark 2.2, both the existence and the uniqueness of the Cauchy problem take place in a neighborhood of the initial moment t_0. However, we expect the uniqueness to have a global nature, that is, if two solutions $\boldsymbol{x} = \boldsymbol{x}(t)$ and $\boldsymbol{y} = \boldsymbol{y}(t)$ of (2.39) are equal at a point t_0, then they should coincide on the common interval of existence. (Their equality on a neighborhood of t_0 follows from the local uniqueness result.)

The next theorem, which is known in the literature as the *global uniqueness theorem*, states that global uniqueness holds under the assumptions of Theorem 2.6.

Theorem 2.7 *Assume that $\boldsymbol{f} : \Omega \to \mathbb{R}^n$ satisfies the assumptions in Theorem 2.6. If $\boldsymbol{x}, \boldsymbol{y}$ are two solutions of (2.39) defined on the open intervals I and J, respectively, and if $\boldsymbol{x}(t_0) = \boldsymbol{y}(t_0)$ for some $t_0 \in I \cap J$, then $\boldsymbol{x}(t) = \boldsymbol{y}(t)$, $\forall t \in I \cap J$.*

Proof Let $(t_1, t_2) = I \cap J$. We will prove that $\boldsymbol{x}(t) = \boldsymbol{y}(t), \forall t \in [t_0, t_2)$. The equality to the left of t_0 is proved in a similar fashion. Let

$$\mathcal{T} := \{ \tau \in [t_0, t_2);\ \ \boldsymbol{x}(t) = \boldsymbol{y}(t);\ \ \forall t \in [t_0, \tau] \}.$$

Then $\mathcal{T} \neq \emptyset$ and we set $T := \sup \mathcal{T}$. We claim that $T = t_2$.

To prove the claim, we argue by contradiction. Assume that $T < t_2$. Then $\boldsymbol{x}(t) = \boldsymbol{y}(t)$, $\forall t \in [t_0, T]$, and since $\boldsymbol{x}(t)$ and $\boldsymbol{y}(t)$ are both solutions of (2.39), we deduce from Theorem 2.6 that there exists a $\varepsilon > 0$ such that $\boldsymbol{x}(t) = \boldsymbol{y}(t)$, $\forall t \in [T, T + \varepsilon]$. This contradicts the maximality of T and concludes the proof of the theorem. □

Remark 2.5 If the function $\boldsymbol{f} : \Omega \to \mathbb{R}^n$ is of class C^{k-1} on the domain Ω, then, obviously, the local solution of system (2.39) is of class C^k on the interval it is defined. Moreover, if $\boldsymbol{f}$ is real analytic on Ω, that is, it is C^∞, and the Taylor series of $\boldsymbol{f}$ at any point $(t_0, \boldsymbol{x}_0) \in \Omega$ converges to $\boldsymbol{f}$ in a neighborhood of that point, then any solution $\boldsymbol{x}$ of (2.39) is also real analytic.

This follows by direct computation from equations (2.39), and, using the fact that a real function $g = g(x_1, \dots, x_m)$ defined on a domain D of $\mathbb{R}^m$ is real analytic if and only if, for any compact set $K \subset D$ and any positive integer k, there exists a positive constant $M(k)$ such that for any multi-index $\alpha = (\alpha_1, \dots, \alpha_m) \in \mathbb{Z}^m_{\geq 0}$ we have

$$\left| \frac{\partial^{|\alpha|} f(\boldsymbol{x})}{\partial^{\alpha_1}_{x_1} \cdots \partial^{\alpha_m}_{x_m}} \right| \leq M\big(|\alpha|\big)^{|\alpha|} \alpha!, \quad \forall \boldsymbol{x} = (x_1, \dots, x_m) \in K,$$

where

$$|\alpha| := \alpha_1 + \cdots + \alpha_m, \quad \alpha! := \alpha_1! \cdots \alpha_m!$$

A solution $\boldsymbol{x} = \varphi(t)$ of (2.39) defined on the interval $I = [a, b]$ is called *extendible* if there exists a solution $\psi(t)$ of (2.39) defined on an interval $J \supsetneq I$ such that $\varphi = \psi$ on I. The solution φ is called *right-extendible* if there exists $b' > b$ and a solution ψ of (2.39), defined on $[a, b']$, such that $\psi = \varphi$ on $[a, b]$. The notion of *left-extendible* solutions is defined analogously. A solution that is not extendible is called *saturated*. In other words, a solution φ defined on an interval I is saturated if I is its maximal domain of existence. Similarly, a solution that is not right-extendible (respectively left-extendible) is called *right-saturated* (respectively *left-saturated*).

Theorem 2.6 implies that a maximal interval on which a saturated solution is defined must be an open interval. If a solution φ is right-saturated, then the interval on which it is defined is open on the right. Similarly, if a solution φ is left-saturated, then the interval on which it is defined is open on the left.

Indeed, if $\varphi : [a, b) \to \mathbb{R}^n$ is a solution of (2.39) defined on an interval that is not open on the left, then Theorem 2.6 implies that there exists a solution $\widetilde{\varphi}(t)$ defined on an interval $[a - \delta, a + \delta]$ satisfying the initial condition $\widetilde{\varphi}(a) = \varphi(a)$. The local uniqueness theorem implies that $\widetilde{\varphi} = \varphi$ on $[a, a + \delta]$ and thus the function

$$\widehat{\varphi}_0(t) = \begin{cases} \varphi(t), & t \in [a, b), \\ \widetilde{\varphi}(t), & t \in [a - \delta, a], \end{cases}$$

is a solution of (2.39) on $[a - \delta, b)$ that extends φ, showing that φ is not left-saturated.

As an illustration, consider the ODE

$$x' = x^2 + 1,$$

with the initial condition $x(t_0) = x_0$. This is a separable ODE and we find that

$$x(t) = \tan\big(t - t_0 + \arctan x_0\big).$$

It follows that, on the right, the maximal existence interval is $[t_0,\ t_0 + \frac{\pi}{2} - \arctan x_0)$, while on the left, the maximal existence interval is $(t_0 - \frac{\pi}{2} -$

$\arctan x_0, t_0]$. Thus, the saturated solution is defined on the interval $(t_0 - \frac{\pi}{2} - \arctan x_0, t_0 + \frac{\pi}{2} - \arctan x_0)$.

Our next result characterizes the right-saturated solutions. In the remainder of this section, we will assume that $\Omega \subset \mathbb{R}^{n+1}$ is an open subset and $\boldsymbol{f} : \Omega \to \mathbb{R}^n$ is a continuous map that is also locally Lipschitz in the variable $\boldsymbol{x} \in \mathbb{R}^n$.

Theorem 2.8 *Let* $\varphi : [t_0, t_1) \to \mathbb{R}^n$ *be a solution to system* (2.39). *Then the following are equivalent.*

(i) *The solution* φ *is right-extendible.*
(ii) *The graph of* φ,

$$\Gamma := \{ (t, \varphi(t)); \quad t \in [t_0, t_1) \},$$

is contained in a compact subset of Ω.

Proof (i) $\Rightarrow$ (ii). Assume that φ is right-extendible. Thus, there exists a solution $\psi(t)$ of (2.39) defined on an interval $[t_0, t_1 + \delta)$, $\delta > 0$, and such that

$$\psi(t) = \varphi(t), \quad \forall t \in [t_0, t_1).$$

In particular, it follows that Γ is contained in $\widehat{\Gamma}$, the graph of the restriction of ψ to $[t_0, t_1]$. Now, observe that $\widehat{\Gamma}$ is a compact subset of Ω because it is the image of the compact interval $[t_0, t_1]$ via the continuous map $t \mapsto (t, \psi(t))$.

(ii) $\Rightarrow$ (i) Assume that $\Gamma \subset K$, where K is a compact subset of Ω. We will prove that $\varphi(t)$ can be extended to a solution of (2.39) on an interval of the form $[t_0, t_1 + \delta]$, for some $\delta > 0$.

Since $\varphi(t)$ is a solution, we have

$$\varphi(t) = \varphi(t_0) + \int_{t_0}^{t} \boldsymbol{f}\big(s, \varphi(s)\big)ds, \quad \forall t \in [t_0, t_1).$$

We deduce that

$$\|\varphi(t) - \varphi(t')\| \le \left| \int_{t'}^{t} \left\| \boldsymbol{f}\big(s, \varphi(s)\big) \right\| ds \right| \le M_K|t - t'|, \quad \forall t, t' \in [t_0, t_1),$$

where $M_K := \sup_{(s,\boldsymbol{x})\in K} \|\boldsymbol{f}(s, \boldsymbol{x})\|$. Cauchy's characterization of convergence now shows that $\varphi(t)$ has a (finite) limit as $t \nearrow t_1$ and we set

$$\varphi(t_1) := \lim_{t \nearrow t_1} \varphi(t).$$

We have thus extended φ to a continuous function on $[t_0, t_1]$ that we continue to denote by φ. The continuity of $\boldsymbol{f}$ implies that

$$\varphi'(t_1 - 0) = \lim_{t \nearrow t_1} \varphi'(t) = \lim_{t \nearrow t_1} \boldsymbol{f}(t, \varphi(t)) = \boldsymbol{f}(t_1, \varphi(t_1)). \tag{2.42}$$

On the other hand, according to Theorem 2.6, there exists a solution $\psi(t)$ of (2.39) defined on an interval $[t_1-\delta, t_1+\delta]$ and satisfying the initial condition $\psi(t_1) = \varphi(t_1)$. Consider the function

$$\widetilde{\varphi}(t) = \begin{cases} \varphi(t), & t \in [t_0, t_1], \\ \psi(t), & t \in (t_1, t_1+\delta]. \end{cases}$$

Obviously,

$$\widetilde{\varphi}'(t_1+0) = \psi'(t_1) = f(t_1, \psi(t_1)) = f(t_1, \varphi(t_1)) \overset{(2.42)}{=} \varphi'(t_1-0).$$

This proves that $\widetilde{\varphi}$ is C^1, and satisfies the differential equation (2.39). Clearly, $\widetilde{\varphi}$ extends φ to the right. □

The next result shows that any solution can be extended to a saturated solution.

Theorem 2.9 *Any solution φ of* (2.39) *admits a unique extension to a saturated solution.*

Proof The uniqueness is a consequence of Theorem 2.7 on global uniqueness. To prove the extendibility to a saturated solution, we will limit ourselves to proving the extendibility to a right-saturated solution.

We denote by $\mathcal{A}$ the set of all solutions ψ of (2.39) that extend φ to the right. The set $\mathcal{A}$ is totally ordered by the inclusion of the domains of definition of the solutions ψ and, as such, the set $\mathcal{A}$ has an upper bound, $\widetilde{\varphi}$. This is a right-saturated solution of (2.39). □

We will next investigate the behavior of the saturated solutions of (2.39) in a neighborhood of the boundary $\partial\Omega$ of the domain Ω where (2.39) is defined. For simplicity, we only discuss the case of right-saturated solutions. The case of left-saturated solutions is identical.

Theorem 2.10 *Let $\varphi(t)$ be a right-saturated solution of* (2.39) *defined on the interval $[t_0, T)$. Then any limit point as $t \nearrow T$ of the graph*

$$\Gamma := \big\{ (t, \varphi(t));\ \ t_0 \le t < T \big\}$$

is either the point at infinity of $\mathbb{R}^{n+1}$, or a point on $\partial\Omega$.

Proof The theorem states that, if (τ_ν) is a sequence in $[t_0, T)$ such that the limit $\lim_{\nu\to\infty}(\tau_\nu, \varphi(\tau_\nu))$ exists, then

(i) either $T = \infty$,

(ii) or $T < \infty$, $\lim_{\nu\to\infty} \|\varphi(\tau_\nu)\| = \infty$,

(iii) or $T < \infty$, $\boldsymbol{x}^* = \lim_{\nu\to\infty} \varphi(\tau_\nu) \in \mathbb{R}^n$ and $(T, \boldsymbol{x}^*) \in \partial\Omega$.

We argue by contradiction. Assume that all three options are violated. Since (i), (ii) do not hold, we deduce that $T < \infty$ and that the limit $\lim_{\nu\to\infty} \varphi(\tau_\nu)$ exists and is

Fig. 2.1 The behavior of a right-saturated solution

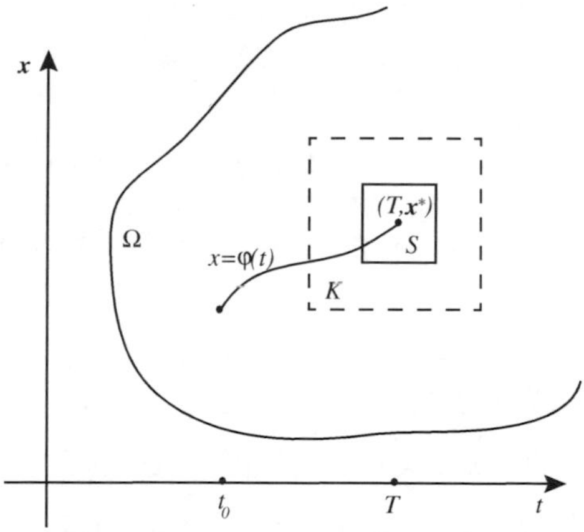

a point $x^* \in \mathbb{R}^n$. Since (iii) is also violated, we deduce that $(T, x^*) \in \Omega$. Thus, for $r > 0$ sufficiently small, the closed ball

$$S := \{ (t, x) \in \mathbb{R}^{n+1}; \quad |t - T| \le r, \quad \|x - x^*\| \le r \}$$

is contained in Ω; see Fig. 2.1.

If $\eta := \text{dist}(S, \partial\Omega) > 0$, we deduce that for any $(s_0, y_0) \in S$ the parallelepiped

$$\Delta := \left\{ (t, x) \in \mathbb{R}^{n+1}; \quad |t - s_0| \le \frac{\eta}{4}, \quad \|x - y_0\| \le \frac{\eta}{4} \right\} \tag{2.43}$$

is contained in the compact subset of Ω,

$$K = \left\{ (t, x) \in \mathbb{R}^{n+1}; \quad |t - T| \le r + \frac{\eta}{2}, \quad \|x - x^*\| \le r + \frac{\eta}{2} \right\}.$$

(See Fig. 2.1.) We set

$$\delta := \min\left\{ \frac{\eta}{4}, \frac{\eta}{4M} \right\}, \quad M := \sup_{(t,x) \in K} \|f(t, x)\|.$$

Appealing to the existence and uniqueness theorem (Theorem 2.3), where Δ is defined in (2.43), it follows that, for any $(s_0, y_0) \in S$, there exists a unique solution $\psi_{s_0, y_0}(t)$ of (2.39) defined on the interval $[s_0 - \delta, s_0 + \delta]$ and satisfying the initial condition $\psi(s_0) = y_0$.

Fix ν sufficiently large so that

$$(\tau_\nu, \varphi(\tau_\nu)) \in S \text{ and } |\tau_\nu - T| \le \frac{\delta}{2},$$

and define $y_\nu := \varphi(\tau_\nu)$,

$$\widetilde{\varphi}(t) := \begin{cases} \varphi(t), & t_0 \leq t \leq \tau_\nu, \\ \psi_{\tau_\nu, y_\nu}(t), & \tau_\nu < t \leq \tau_\nu + \delta. \end{cases}$$

Then $\widetilde{\varphi}(t)$ is a solution of (2.39) defined on the interval $[t_0, \tau_\nu + \delta]$. This interval strictly contains the interval $[t_0, T]$ and $\widetilde{\varphi} = \varphi$ on $[t_0, T)$. This contradicts our assumption that φ is a right-saturated solution, and completes the proof of Theorem 2.10. □

Theorem 2.11 *Let* $\Omega = \mathbb{R}^{n+1}$ *and* $\varphi(t)$ *be a right-saturated solution of* (2.39) *defined on* $[0, T)$. *Then only the following two options are possible:*

(i) either $T = \infty$,
(ii) or $T < \infty$ and $\lim_{t \nearrow T} \|\varphi(t)\| = \infty$.

Proof From Theorem 2.10 it follows that any limit point as $t \nearrow T$ on the graph Γ of φ is the point at infinity. If $T < \infty$, then necessarily

$$\lim_{t \nearrow T} \|\varphi(t)\| = \infty. \qquad \square$$

Theorems 2.10 and 2.11 are useful in determining the maximal existence interval of a solution. Loosely speaking, Theorem 2.11 states that a solution φ is either defined on the whole positive semi-axis, or it "blows up" in finite time. This phenomenon is commonly referred to as the finite-time *blowup* phenomenon.

To illustrate Theorem 2.11, we depict in Fig. 2.2 the graph of the saturated solution of the Cauchy problem

$$x' = x^2 - 1, \quad x(0) = 2.$$

Its maximal existence interval on the right is $[0, T)$, $T = \frac{1}{2} \log 3$.

In the following examples, we describe other applications of these theorems.

Example 2.1 Consider the scalar ODE

$$\boldsymbol{x}' = \boldsymbol{f}(\boldsymbol{x}), \tag{2.44}$$

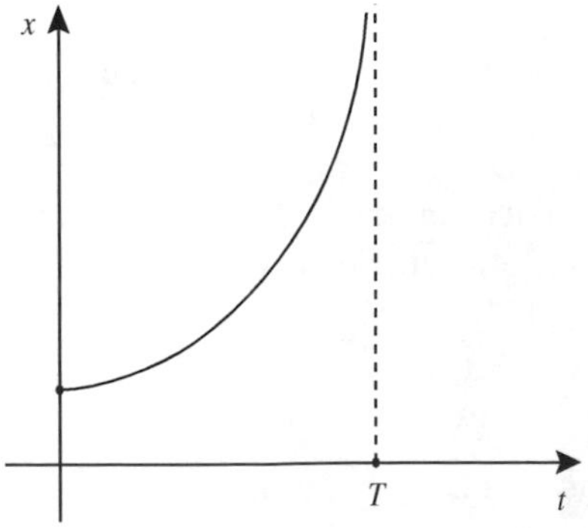

Fig. 2.2 A finite-time blowup

where $\boldsymbol{f} : \mathbb{R}^n \to \mathbb{R}^n$ is locally Lipschitz and satisfies

$$(\boldsymbol{x}, \boldsymbol{f}(\boldsymbol{x})) \leq \gamma_1 \|\boldsymbol{x}\|_{\mathbf{e}}^2 + \gamma_2, \quad \forall \boldsymbol{x} \in \mathbb{R}^n, \tag{2.45}$$

where $\gamma_1, \gamma_2 \in \mathbb{R}$. (Here $(-, -)$ is the Euclidean scalar product on $\mathbb{R}^n$ and $\| - \|_{\mathbf{e}}$ is the Euclidean norm.) According to the existence and uniqueness theorem, for any $(t_0, \boldsymbol{x}_0) \in \mathbb{R}^2$ there exists a unique solution $\varphi(t) = \boldsymbol{x}(t; t_0, \boldsymbol{x}_0)$ of (2.44) satisfying $\varphi(t_0) = \boldsymbol{x}_0$ and defined on a maximal interval $[t_0, T)$. We want to prove that under the above assumptions we have $T = \infty$.

To show this, we multiply scalarly both sides of (2.44) by $\varphi(t)$. Using (2.45), we deduce that

$$\frac{1}{2}\frac{d}{dt}\|\varphi(t)\|_{\mathbf{e}}^2 = (\varphi(t), \varphi'(t)) = (\boldsymbol{f}(\varphi(t)), \varphi(t)) \leq \gamma_1 \|\varphi(t)\|_{\mathbf{e}}^2 + \gamma_2, \ \forall t \in [t_0, T),$$

and, therefore,

$$\|\varphi(t)\|_{\mathbf{e}}^2 \leq \|\varphi(t_0)\|_{\mathbf{e}}^2 + \gamma_1 \int_0^t \|\varphi(s)\|_{\mathbf{e}}^2 \, ds + \gamma_2 T, \quad \forall t \in [t_0, T).$$

Then, by Gronwall's lemma (Lemma 1.1), we get

$$\|\varphi(t)\|_{\mathbf{e}}^2 \leq (\|\varphi(t_0)\|_{\mathbf{e}}^2 + \gamma_2 T) \exp(\gamma_1 T), \quad \forall\, t \in (0, T).$$

Thus, the solution $\varphi(t)$ is bounded, and so there is no blowup, $T = \infty$.

It should be noted that, in particular, condition (2.45) holds if f is globally Lipschitz on $\mathbb{R}^n$.

Example 2.2 Consider the Riccati equation

$$x' = a(t)x + b(t)x^2 + c(t), \tag{2.46}$$

where $a, b, c : [0, \infty) \to \mathbb{R}$ are continuous functions. We associate with (2.46) the Cauchy condition

$$x(t_0) = x_0, \tag{2.47}$$

where $t_0 = 0$. We will prove the following result.

If $x_0 \geq 0$ *and*

$$b(t) \leq 0, \quad c(t) \geq 0, \quad \forall t \geq 0,$$

then the Cauchy problem (2.46)–(2.47) *admits a unique solution* $x = \varphi(t)$ *defined on the semi-axis* $[t_0, \infty)$. *Moreover,* $\varphi(t) \geq 0$, $\forall t \geq t_0$.

We begin by proving the result under the stronger assumption

$$c(t) > 0, \quad \forall t \geq 0.$$

Let $\varphi(t)$ denote the right-saturated solution of (2.46) and (2.47). It is defined on a maximal interval $[t_0, T)$. We will first prove that

$$\varphi(t) \geq 0, \quad \forall t \in [t_0, T). \tag{2.48}$$

Note that, if $x_0 = 0$, then $\varphi'(t_0) = c(t_0) > 0$, so $\varphi(t) > 0$ for t in a small interval $[t_0, t_0 + \delta]$, $\delta > 0$. This reduces the problem to the case when the initial condition is positive. Assume, therefore, that $x_0 > 0$. There exists a maximal interval $[t_0, T_1) \subset [t_0, T)$ on which $\varphi(t)$ is nonnegative. Clearly, either $T_1 = T$, or $T_1 < T$ and $\varphi(T_1) = 0$. If $T_1 < T$, then arguing as above we can extend φ past T_1 while keeping it nonnegative. This contradicts the maximality of T_1, thus proving (2.48).

To prove that $T = \infty$, we will rely on Theorem 2.11 and we will show that $\varphi(t)$ cannot blow up in finite time. Using the equality

$$\varphi'(t) = a(t)\varphi(t) + b(t)\varphi(t)^2 + c(t), \quad \forall t \in [t_0, T)$$

and the inequalities $b(t) \leq 0$, $\varphi(t) \geq 0$, we deduce that

$$|\varphi(t)| = \varphi(t) \leq \underbrace{\varphi(t_0) + \int_{t_0}^{t} c(s)ds}_{=:\beta(t)} + \int_{t_0}^{t} |a(s)|\,|\varphi(s)|ds.$$

We can invoke Gronwall's lemma to conclude that

$$|\varphi(t)| \leq \beta(t) + \int_{t_0}^{t} \beta(s)|a(s)| \exp\left(\int_{s}^{t} |a(\tau)|d\tau\right) ds.$$

The function in the right-hand-side of the above inequality is continuous on $[t_0, \infty)$, showing that $\varphi(t)$ cannot blow up in finite time. Hence $T = \infty$.

To deal with the general case, when $c(t) \geq 0$, $\forall t \geq 0$, we consider the equation

$$x' = a(t)x + b(t)x^2 + c(t) + \varepsilon, \tag{2.49}$$

where $\varepsilon > 0$. According to the results proven so far, this equation has a unique solution $x_\varepsilon(t; t_0, x_0)$ satisfying (2.47) and defined on $[t_0, \infty)$.

Denote by $x(t; t_0, x_0)$ the right-saturated solution of (2.46) and (2.47), defined on a maximal interval $[t_0, T)$. According to the forthcoming Theorem 2.15, we have

$$\lim_{\varepsilon \searrow 0} x_\varepsilon(t; t_0, x_0) = x(t; t_0, x_0), \quad \forall t \in [t_0, T).$$

We conclude that $x(t; t_0, x_0) \geq 0$, $\forall t \in [t_0, T)$. Using Gronwall's lemma as before, we deduce that $x(t; t_0, x_0)$ cannot blow up in finite time, and thus $T = \infty$.

Example 2.3 Let A be a real $n \times n$ matrix and Q, X_0 be two real symmetric, nonnegative definite $n \times n$ matrices. We recall that a symmetric $n \times n$ matrix S is called *nonnegative definite* if

$$(S\boldsymbol{v}, \boldsymbol{v}) \geq 0, \quad \forall \boldsymbol{v} \in \mathbb{R}^n,$$

where $(-, -)$ denotes the canonical scalar product on $\mathbb{R}^n$. The symmetric matrix S is called *positive definite* if

$$(S\boldsymbol{v}, \boldsymbol{v}) > 0, \quad \forall \boldsymbol{v} \in \mathbb{R}^n \setminus \{0\}.$$

We denote by A^* the adjoint (transpose) of S and we consider the matrix differential equation

$$X'(t) + A^*X(t) + X(t)A + X(t)^2 = Q, \tag{2.50}$$

together with the initial condition

$$X(t_0) = X_0. \tag{2.51}$$

By a solution of Eq. (2.50), we understand a matrix-valued map

$$X : I \to \mathbb{R}^{n^2}, \quad t \mapsto X(t) = \big(x_{ij}(t)\big)_{1 \leq i,j \leq n}$$

of class C^1 that satisfies (2.50) everywhere on I. Thus (2.50) is a system of ODEs involving n^2 unknown functions. When $n = 1$, Eq. (2.50) reduces to (2.46). Equation (2.50) is called the *matrix-valued Riccati type* equation and it plays an important role in the theory of control systems with quadratic cost functions. In such problems, one is interested in finding global solutions $X(t)$ of (2.50) such that $X(t)$ is symmetric and nonnegative definite for any t. (See Eq. (5.114).)

Theorem 2.12 *Under the above assumptions, the Cauchy problem* (2.50) and (2.51) *admits a unique solution $X = X(t)$ defined on the semi-axis* $[t_0, \infty)$. *Moreover, $X(t)$ is symmetric and nonnegative definite for any $t \geq t_0$.*

Proof From Theorem 2.3, we deduce the existence and uniqueness of a right-saturated solution of this Cauchy problem defined on a maximal interval $[t_0, T)$. Taking the adjoints of both sides of (2.50) and using the fact that X_0 and Q are symmetric matrices, we deduce that $X^*(t)$ is also a solution of the same Cauchy problem (2.50) and (2.51). This proves that $X(t) = X^*(t)$, that is, $X(t)$ is symmetric for any $t \in [t_0, T)$.

Let us prove that

(i) the matrix $X(t)$ is also nonnegative definite for any $t \in [t_0, T)$, and
(ii) $T = \infty$.

We distinguish two cases.

1. The matrix Q is positive definite. Recall that $X(t_0) = X_0$ is nonnegative definite. We set

$$T' := \sup\{\tau \in [t_0, T);\quad X(t) \geq 0,\quad \forall t \in [t_0, \tau)\}.$$

We have to prove that $T' = T$. If $T' < T$, then $X(T')$ is nonnegative definite and there exist sequences (t_k) in (T', T) and $(\boldsymbol{v}_k)$ in $\mathbb{R}^n$ with the following properties

- $\lim_{k\to\infty} t_k = T'$.
- $\|\boldsymbol{v}_k\|_{\mathbf{e}} = 1$, $\big(X(t_k)\boldsymbol{v}_k, \boldsymbol{v}_k\big) < 0$, $\forall k$, where $\|-\|_{\mathbf{e}}$ is the standard Euclidean norm on $\mathbb{R}^n$.
- $\exists \boldsymbol{v}^* \in \mathbb{R}^n$ such that $\boldsymbol{v}^* = \lim_{k\to\infty} \boldsymbol{v}_k$, $X(T')\boldsymbol{v}^* = 0$.

For each $\boldsymbol{v} \in \mathbb{R}^n$, we define the functions

$$\varphi_{\boldsymbol{v}},\quad \psi_{\boldsymbol{v}} : [t_0, T) \to \mathbb{R},\quad \varphi_{\boldsymbol{v}}(t) = \big(X(t)\boldsymbol{v}, \boldsymbol{v}\big),\quad \psi_{\boldsymbol{v}}(t) = \big(X(t)\boldsymbol{v}, A\boldsymbol{v}\big).$$

Since $X(t)$ is symmetric, from (2.50) we see that $\varphi_{\boldsymbol{v}}(t)$ satisfies the ODE

$$\varphi_{\boldsymbol{v}}'(t) = -2\psi_{\boldsymbol{v}}(t) - \|X(t)\boldsymbol{v}\|_{\mathbf{e}}^2 + (Q\boldsymbol{v}, \boldsymbol{v}). \tag{2.52}$$

Moreover, we have

$$\varphi_{\boldsymbol{v}^*}(T') = \psi_{\boldsymbol{v}^*}(T') = 0,$$

and

$$\varphi_{\boldsymbol{v}^*}'(T') = (Q\boldsymbol{v}^*, \boldsymbol{v}^*) > 0. \tag{2.53}$$

Using the mean value theorem, we deduce that for any k there exists an $s_k \in (T', t_k)$ such that

$$\varphi_{\boldsymbol{v}_k}'(s_k) = \frac{\varphi_{\boldsymbol{v}_k}(t_k) - \varphi_{\boldsymbol{v}_k}(T')}{t_k - T'}.$$

We note that, by definition of T', $\varphi_{\boldsymbol{v}_k}(T') \geq 0$. Since $\varphi_{\boldsymbol{v}_k}(t_k) < 0$, we deduce that $\varphi_{\boldsymbol{v}_k}'(s_k) > 0$. Observing that

$$\lim_{k\to\infty} \varphi_{\boldsymbol{v}_k}'(s_k) = \lim_{k\to\infty} \big(X'(s_k)\boldsymbol{v}_k, \boldsymbol{v}_k\big) = \big(X'(T')\boldsymbol{v}^*, \boldsymbol{v}^*\big) = \varphi_{\boldsymbol{v}^*}'(T'),$$

we deduce that $\varphi_{\boldsymbol{v}^*}'(T') \leq 0$.

This contradicts (2.53) and proves that $X(t) \geq 0$, $\forall t \in [t_0, T')$.

According to Theorem 2.11, to prove that $T = \infty$ it suffices to show that for any $\boldsymbol{v} \in \mathbb{R}^n$ there exists a continuous function $f_{\boldsymbol{v}} : [t_0, \infty) \to \mathbb{R}$ such that

$$\varphi_{\boldsymbol{v}}(t) \leq f(t),\quad \forall t \in [t_0, T).$$

Fix $\boldsymbol{v} \in \mathbb{R}^n$. Using the Cauchy–Schwarz inequality (Lemma A.4),

$$|\psi_{\boldsymbol{v}}(t)| = \big|\big(X(t)\boldsymbol{v}, A\boldsymbol{v}\big)\big| \leq \underbrace{\|X(t)\boldsymbol{v}\|_{\mathbf{e}}}_{=:g_{\boldsymbol{v}}(t)} \cdot \underbrace{\|A\boldsymbol{v}\|_{\mathbf{e}}}_{=:C_{\boldsymbol{v}}}.$$

Using this in (2.52), we get

$$\varphi_v'(t) \le 2C_{\boldsymbol{v}} g_{\boldsymbol{v}}(t) - (g_{\boldsymbol{v}}(t))^2 + (Q\boldsymbol{v}, \boldsymbol{v})$$

and, therefore,

$$\begin{aligned} \varphi_{\boldsymbol{v}}(t) \le f_{\boldsymbol{v}}(t) &:= \varphi_{\boldsymbol{v}}(t_0) + (t - t_0)(Q\boldsymbol{v}, \boldsymbol{v}) \\ &\quad + \int_{t_0}^{t} \Big(2C_{\boldsymbol{v}} g_{\boldsymbol{v}}(s) - (g_{\boldsymbol{v}}(s))^2 \Big) ds, \end{aligned} \tag{2.54}$$

$\forall t \in [t_0, T)$. This proves that $T = \infty$.

2. The matrix Q is only nonnegative definite. For any $\varepsilon > 0$, we set $Q_\varepsilon := Q + \varepsilon \mathbb{1}_n$, where $\mathbb{1}_n$ denotes the identity $n \times n$ matrix. Denote by $X_\varepsilon(t)$ the right-saturated solution of the Cauchy problem

$$X'(t) + A^* X(t) + X(t)A + X(t)^2 = Q_\varepsilon, \quad X_\varepsilon(t_0) = X_0.$$

According to the previous considerations, $X_\varepsilon(t)$ is defined on $[t_0, \infty)$ and it is nonnegative definite on this interval. Moreover, for any $\boldsymbol{v} \in \mathbb{R}^n$, any $\varepsilon > 0$ and any $t \ge t_0$, we have

$$\begin{aligned} (X_\varepsilon(t)\boldsymbol{v}, \boldsymbol{v}) \le f_{\boldsymbol{v}}^\varepsilon(t) &:= (X_0\boldsymbol{v}, \boldsymbol{v}) + (t - t_0)(Q\boldsymbol{v}, \boldsymbol{v}) \\ &\quad + \int_{t_0}^{t} \Big(2C_{\boldsymbol{v}} g_{\boldsymbol{v}}^\varepsilon(s) - g_{\boldsymbol{v}}(s)^2 \Big) ds, \end{aligned} \tag{2.55}$$

$$g_{\boldsymbol{v}}^\varepsilon(t) := \| X_\varepsilon(t)\boldsymbol{v} \|_{\mathbf{e}}.$$

From Theorem 2.15, we deduce that

$$\lim_{\varepsilon \searrow 0} X_\varepsilon(t) = X(t), \quad \forall t \in [t_0, T).$$

If we now let $\varepsilon \to 0$ in (2.55), we deduce that

$$(X_\varepsilon(t)\boldsymbol{v}, \boldsymbol{v}) \le f_{\boldsymbol{v}}(t) \;\; \forall t \in [t_0, T),$$

where $f_{\boldsymbol{v}}(t)$ is defined in (2.54). This implies that $T = \infty$. □

Example 2.4 (*Dissipative systems of ODEs*) As a final application, we consider *dissipative, autonomous differential systems*, that is, systems of ordinary differential equations of the form

$$\boldsymbol{x}' = \boldsymbol{f}(\boldsymbol{x}), \tag{2.56}$$

where $\boldsymbol{f} : \mathbb{R}^n \to \mathbb{R}^n$ is a continuous map satisfying the *dissipativity* condition

$$\big(\boldsymbol{f}(\boldsymbol{x}) - \boldsymbol{f}(\boldsymbol{y}), \boldsymbol{x} - \boldsymbol{y} \big) \le 0, \quad \forall \boldsymbol{x}, \boldsymbol{y} \in \mathbb{R}^n, \tag{2.57}$$

where, as usual, $(-,-)$ denotes the canonical Euclidean scalar product on $\mathbb{R}^n$. We associate with (2.56) the initial condition

$$x(t_0) = x_0, \tag{2.58}$$

where (t_0, x_0) is a given point in $\mathbb{R}^{n+1}$.

Mathematical models of a large class of physical phenomena, such as diffusion, lead to dissipative differential systems. In the case $n = 1$, the monotonicity condition (2.57) is equivalent to the requirement that f be monotonically nonincreasing. For dissipative systems, we have the following interesting existence and uniqueness result.

Theorem 2.13 *If the continuous map* $f : \mathbb{R}^n \to \mathbb{R}^n$ *is dissipative, then for any* $(t_0, x_0) \in \mathbb{R}^{n+1}$ *the Cauchy problem* (2.56) and (2.57) *admits a unique solution* $x = x(t; t_0, x_0)$ *defined on* $[t_0, \infty)$. *Moreover, the map*

$$S : [0, \infty) \times \mathbb{R}^n \to \mathbb{R}^n, \quad (t, x_0) \mapsto S(t)x_0 := x(t; 0, x_0),$$

satisfies the following properties.

$$S(0)x_0 = x_0, \quad \forall x_0 \in \mathbb{R}^n, \tag{2.59}$$

$$S(t+s)x_0 = S(t)S(s)x_0, \quad \forall x_0 \in \mathbb{R}^n, \quad t, s \geq 0, \tag{2.60}$$

$$\|S(t)x_0 - S(t)y_0\|_{\mathbf{e}} \leq \|x_0 - y_0\|_{\mathbf{e}}, \quad \forall t \geq 0, \quad x_0, y_0 \in \mathbb{R}^n. \tag{2.61}$$

Proof According to Peano's theorem, for any $(t_0, x_0) \in \mathbb{R}^n$ there exists a solution $x = \varphi(t)$ to the Cauchy problem (2.56) and (2.57) defined on a maximal interval $[t_0, T)$. To prove its uniqueness, we argue by contradiction and assume that this Cauchy problem admits another solution $x = \widetilde{\varphi}(t)$. On their common domain of existence $[t_0, t_1)$, the functions φ and $\widetilde{\varphi}$ satisfy the differential system

$$\big(\varphi(t) - \widetilde{\varphi}(t)\big)' = f\big(\varphi(t)\big) - f\big(\widetilde{\varphi}(t)\big). \tag{2.62}$$

Taking the scalar product of both sides of (2.62) with $\varphi(t) - \widetilde{\varphi}(t)$, we get

$$\begin{aligned} \frac{1}{2}\frac{d}{dt}\|\varphi(t) - \widetilde{\varphi}(t)\|_{\mathbf{e}}^2 &\overset{\text{Lemma A.6}}{=} \Big(\big(\varphi(t) - \widetilde{\varphi}(t)\big)', \ \varphi(t) - \widetilde{\varphi}(t)\Big) \\ &= \Big(f\big(\varphi(t)\big) - f\big(\widetilde{\varphi}(t)\big), \varphi(t) - \widetilde{\varphi}(t)\Big) \overset{(2.57)}{\leq} 0, \quad \forall t \in [t_0, t_1). \end{aligned} \tag{2.63}$$

Thus

$$\|\varphi(t) - \widetilde{\varphi}(t)\|_{\mathbf{e}}^2 \leq \|\varphi(t_0) - \widetilde{\varphi}(t_0)\|_{\mathbf{e}}^2, \quad \forall t \in [t_0, t_1). \tag{2.64}$$

This proves that $\varphi = \widetilde{\varphi}$ on $[t_0, t_1)$ since $\widetilde{\varphi}(t_0) = \varphi(t_0)$.

To prove that φ is defined on the entire semi-axis $[t_0, \infty)$ we first prove that it is bounded on $[t_0, T)$. To achieve this, we take the scalar product of

$$\varphi'(t) = \boldsymbol{f}(\varphi(t))$$

with $\varphi(t)$ and we deduce that

$$\begin{aligned}\frac{1}{2}\frac{d}{dt}\|\varphi(t)\|_{\mathbf{e}}^2 &= (\boldsymbol{f}(\varphi(t)), \varphi(t)) \\ &= \big(\boldsymbol{f}(\varphi(t)) - \boldsymbol{f}(0), \varphi(t)\big) + \big(\boldsymbol{f}(0), \varphi(t)\big) \\ &\overset{(2.57)}{\leq} \|\boldsymbol{f}(0)\|_{\mathbf{e}} \cdot \|\varphi(t)\|_{\mathbf{e}}, \ \forall t \in [t_0, T).\end{aligned}$$

Integrating this inequality on $[t_0, t]$ and setting $u(t) := \|\varphi(t)\|_{\mathbf{e}}$, $C = \|\boldsymbol{f}(0)\|_{\mathbf{e}}$, we deduce that

$$\frac{1}{2}u(t)^2 \leq \frac{1}{2}\|\boldsymbol{x}_0\|_{\mathbf{e}}^2 + C\int_{t_0}^{t} u(s)ds, \ \ \forall t \in [t_0, T).$$

From Proposition 1.2 we deduce that

$$\|\varphi(t)\|_{\mathbf{e}} = u(t) \leq \|\boldsymbol{x}_0\|_{\mathbf{e}} + \|\boldsymbol{f}(0)\|_{\mathbf{e}}(t - t_0), \ \ \forall t \in [t_0, T). \tag{2.65}$$

Since we have not assumed that the function $\boldsymbol{f}$ is locally Lipschitz, we cannot invoke Theorems 2.10 or 2.11 directly. However, inequality (2.65) implies in a similar fashion the equality $T = \infty$. Here are the details.

We argue by contradiction and we assume that $T < \infty$. Inequality (2.65) implies that there exists an increasing sequence (t_k) and $\boldsymbol{v} \in \mathbb{R}^n$ such that

$$\lim_{k\to\infty} t_k = T, \ \ \lim_{k\to\infty} \varphi(t_k) = \boldsymbol{v}.$$

According to the facts established so far, there exists a solution ψ of (2.56) defined on $[T - \delta, T + \delta]$ and satisfying the initial condition $\psi(T) = \boldsymbol{v}$.

On the interval $[T - \delta, T)$ we have

$$\varphi'(t) - \psi'(t) = \boldsymbol{f}(\varphi(t)) - \boldsymbol{f}(\psi(t)).$$

Taking the scalar product of this equality with $\varphi(t) - \psi(t)$ and using the dissipativity condition (2.57), we deduce as before that

$$\frac{1}{2}\frac{d}{dt}\|\varphi(t) - \psi(t)\|_{\mathbf{e}}^2 \leq 0, \ \ \forall t \in [T - \delta, T).$$

Hence

$$\|\varphi(t) - \psi(t)\|_{\mathbf{e}}^2 \leq \|\varphi(t_k) - \psi(t_k)\|_{\mathbf{e}}^2, \ \ \forall t \in [T - \delta, t_k].$$

Since $\lim_{k\to\infty} \|\varphi(t_k) - \psi(t_k)\|_e = 0$, we conclude that $\varphi = \psi$ on $[T - \delta, T)$. In other words, ψ is a proper extension of the solution φ. This contradicts the maximality of the interval $[t_0, T)$. Thus $T = \infty$.

To prove (2.60), we observe that both functions

$$\boldsymbol{y}_1(t) = S(t+s)\boldsymbol{x}_0 \quad \text{and} \quad \boldsymbol{y}_2(t) = S(t)S(s)\boldsymbol{x}_0$$

satisfy equations (2.56) and have identical values at $t = 0$. The uniqueness of the Cauchy problems for (2.56) now implies that $\boldsymbol{y}_1(t) = \boldsymbol{y}_2(t)$, $\forall t \geq 0$.

Inequality (2.61) now follows from (2.64) where $\widetilde{\varphi}(t) = \boldsymbol{x}(t; 0, \boldsymbol{y}_0)$. □

Remark 2.6 A family of maps $S(t) : \mathbb{R}^n \to \mathbb{R}^n$, $t \geq 0$, satisfying (2.59), (2.60), (2.61) is called a *continuous semigroup of contractions* on the space $\mathbb{R}^n$. The function $\boldsymbol{f} : \mathbb{R}^n \to \mathbb{R}^n$ is called the *generator* of the semigroups $S(t)$.

2.6 Continuous Dependence on Initial Conditions and Parameters

We now return to the differential system (2.39) defined on the open subset $\Omega \subset \mathbb{R}^{n+1}$. We will assume as in the previous section that the function $\boldsymbol{f} : \Omega \to \mathbb{R}^n$ is continuous in the variables $(t, \boldsymbol{x})$, and locally Lipschitz in the variable $\boldsymbol{x}$. Theorem 2.6 shows that for any $(t_0, \boldsymbol{x}_0) \in \Omega$ there exists a unique solution $\boldsymbol{x} = \boldsymbol{x}(t; t_0, \boldsymbol{x}_0)$ of system (2.39) that satisfies the initial condition $\boldsymbol{x}(t_0) = \boldsymbol{x}_0$. The solution $\boldsymbol{x}(t; t_0, \boldsymbol{x}_0)$, which we will assume to be saturated, is defined on an interval typically dependent on the point $(t_0, \boldsymbol{x}_0)$. For simplicity, we will assume the initial moment t_0 to be fixed.

It is reasonable to expect that, as $\boldsymbol{v}$ varies in a neighborhood of $\boldsymbol{x}_0$, the corresponding solution $\boldsymbol{x}(t; t_0, \boldsymbol{v})$ will not stray too far from the solution $\boldsymbol{x}(t; t_0, \boldsymbol{x}_0)$. The next theorem confirms that this is the case, in a rather precise form. To state this result, let us denote by $B(\boldsymbol{x}_0, \eta)$ the ball of center $\boldsymbol{x}_0$ and radius η in $\mathbb{R}^n$, that is,

$$B(\boldsymbol{x}_0, \eta) := \big\{ \boldsymbol{v} \in \mathbb{R}^n;\ \ \|\boldsymbol{v} - \boldsymbol{x}_0\| \leq \eta \big\}.$$

Theorem 2.14 (Continuous dependence on initial data) *Let $[t_0, T)$ be the maximal interval of existence on the right of the solutions $\boldsymbol{x}(t; t_0, \boldsymbol{x}_0)$ of* (2.39). *Then, for any $T' \in [t_0, T)$, there exists an $\eta = \eta(T') > 0$ such that, for any $\boldsymbol{v} \in S(\boldsymbol{x}_0, \eta)$, the solution $\boldsymbol{x}(t; t_0, \boldsymbol{v})$ is defined on the interval $[t_0, T']$. Moreover, the correspondence*

$$B(\boldsymbol{x}_0, \eta) \ni \boldsymbol{v} \mapsto \boldsymbol{x}(t; t_0, \boldsymbol{v}) \in C\big([t_0, T']; \mathbb{R}^n\big)$$

is a continuous map from the ball $S(\boldsymbol{x}_0, \eta)$ to the space $C\big([t_0, T']; \mathbb{R}^n\big)$ of continuous maps from $[t_0, T']$ to $\mathbb{R}^n$. In other words, for any sequence $(\boldsymbol{v}_k)$ in $B(\boldsymbol{x}_0, \eta)$ that

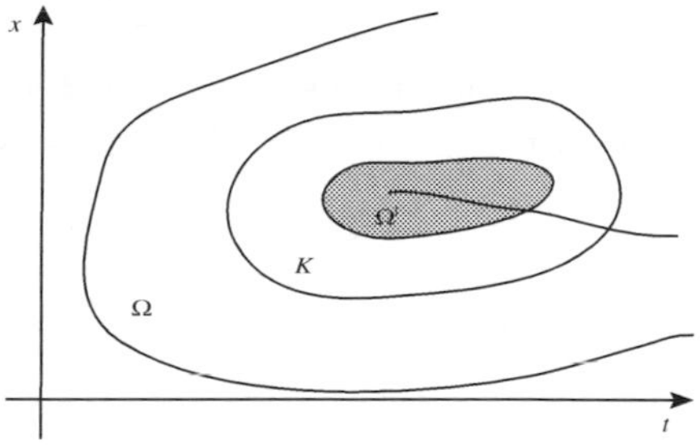

Fig. 2.3 Isolating a compact portion of an integral curve

converges to $\boldsymbol{v} \in B(\boldsymbol{x}_0, \eta)$*, the sequence of functions* $\boldsymbol{x}(t; t_0, \boldsymbol{v}_k)$ *converges uniformly on* $[t_0, T']$ *to* $\boldsymbol{x}(t; t_0, \boldsymbol{v})$.

Proof Fix $T' \in [t_0, T)$. The restriction to $[t_0, T']$ of $\boldsymbol{x}(t; t_0, \boldsymbol{x}_0)$ is continuous and, therefore, the graph of this restriction is compact. We can find an open set Ω' whose closure $\bar{\Omega}'$ is compact and contained in Ω and such that

$$\big\{(t, \boldsymbol{x}(t; t_0, \boldsymbol{x}_0));\ t_0 \leq t \leq T'\big\} \subset \bar{\Omega}',\quad \operatorname{dist}\big(\bar{\Omega}', \partial\Omega\big) =: \delta > 0. \tag{2.66}$$

We denote by K the compact subset of Ω defined by (see Fig. 2.3)

$$K := \Big\{(t, \boldsymbol{x}) \in \Omega;\ \operatorname{dist}((t, \boldsymbol{x}), \bar{\Omega}') \leq \frac{\delta}{2}\Big\}. \tag{2.67}$$

For any $(t_0, \boldsymbol{v}) \in \Omega'$, there exists a maximal $\tilde{T} \in (t_0, T']$ such that the solution $\boldsymbol{x}(t; t_0, \boldsymbol{v})$ exists for all $t \in [t_0, \tilde{T}]$ and $\big\{(t, \boldsymbol{x}(t; t_0, \boldsymbol{v}));\ t_0 \leq t \leq \tilde{T}\big\} \subset K$. On the interval $[t_0, T']$, we have the equality

$$\boldsymbol{x}(t; t_0, \boldsymbol{x}_0) - \boldsymbol{x}(t; t_0, \boldsymbol{v}) = \int_{t_0}^{t} \Big(\boldsymbol{f}\big(s, \boldsymbol{x}(s; t_0, \boldsymbol{x}_0)\big) - \boldsymbol{f}\big(s, \boldsymbol{x}(s; t_0, \boldsymbol{v})\big) \Big) ds.$$

Because the graphs of $\boldsymbol{x}(s; t_0, \boldsymbol{x}_0)$ and $\boldsymbol{x}(s; t_0, \boldsymbol{v})$ over $[t_0, T']$ are contained in the compact set K, the locally Lipschitz assumption implies that there exists a constant $L_K > 0$ such that

$$\big\|\boldsymbol{x}(t; t_0, \boldsymbol{x}_0) - \boldsymbol{x}(t; t_0, \boldsymbol{v})\big\| \leq \|\boldsymbol{x}_0 - \boldsymbol{v}\| + L_K \int_{t_0}^{t} \big\| \boldsymbol{x}(s; t_0, \boldsymbol{x}_0) - \boldsymbol{x}(s; t_0, \boldsymbol{v}) \big\| ds.$$

Gronwall's lemma now implies

$$\big\|\boldsymbol{x}(t; t_0, \boldsymbol{x}_0) - \boldsymbol{x}(t; t_0, \boldsymbol{v})\big\| \leq e^{L_K(t-t_0)} \|\boldsymbol{x}_0 - \boldsymbol{v}\|,\quad \forall t \in [t_0, \tilde{T}]. \tag{2.68}$$

We can now prove that, given $T' \in [t_0, T)$, there exists an $\eta = \eta(T') > 0$ such that, for any $\boldsymbol{v} \in B(\boldsymbol{x}_0, \eta)$,

(a) the solution $\boldsymbol{x}(t; t_0, \boldsymbol{v})$ is defined on $[t_0, T']$, and
(b) the graph of this solution is contained in K.

We argue by contradiction. We can find a sequence $(\boldsymbol{v}_j)_{j\geq 1}$ in $\mathbb{R}^n$ such that

- $\|\boldsymbol{x}_0 - \boldsymbol{v}_j\| \leq \frac{1}{j}, \forall j \geq 1$, and
- the maximal closed interval $[t_0, T_j]$, with the property that the graph of $\boldsymbol{x}(t; t_0, \boldsymbol{v}_j)$ is contained in K, is a subinterval of the half-open interval $[t_0, T')$.

Using (2.68), we deduce that

$$\left\|\boldsymbol{x}(t; t_0, \boldsymbol{x}_0) - \boldsymbol{x}(t; t_0, \boldsymbol{v}_j)\right\| \leq \frac{e^{L_K(t-t_0)}}{j}, \quad \forall t_0 \leq t \leq T_j. \tag{2.69}$$

Thus, if

$$j \geq \frac{\delta e^{L_K(t-t_0)}}{4},$$

then the distance between the graph of $\boldsymbol{x}(t; t_0, \boldsymbol{v}_j)$ and the graph of $\boldsymbol{x}(t; t_0, \boldsymbol{x}_0)$ over $[t_0, T_j]$ is $\leq \frac{\delta}{4}$. Conditions (2.66) and (2.67) imply that

$$\operatorname{dist}((t, \boldsymbol{x}(t; t_0, \boldsymbol{v}_j)), \partial K) \geq \frac{\delta}{4}, \quad \forall t \in [t_0, T_j].$$

We conclude that the function $\boldsymbol{x}(t; t_0, \boldsymbol{v}_j)$ can be extended slightly to the right of T_j as a solution of (2.39) so that its graph continues to be inside K. This violates the maximality of T_j. This proves the existence of $\eta(T')$ with the postulated properties (a) and (b) above.

Consider now two solutions $\boldsymbol{x}(t; t_0, \boldsymbol{u})$, $\boldsymbol{x}(t; t_0, \boldsymbol{v})$, where $\boldsymbol{u}, \boldsymbol{v} \in B(\boldsymbol{x}_0, \eta(T'))$. For $t \in [t_0, T']$, we have

$$\boldsymbol{x}(t; t_0, \boldsymbol{u}) - \boldsymbol{x}(t; t_0, \boldsymbol{v}) = \boldsymbol{u} - \boldsymbol{v} + \int_{t_0}^{t} \Big(\boldsymbol{f}\big(s, \boldsymbol{x}(s; t_0, \boldsymbol{u})\big) - \boldsymbol{f}\big(s, \boldsymbol{x}(s; t_0, \boldsymbol{v})\big) \Big) ds.$$

Using the local Lipschitz condition, we deduce as before that

$$\|\boldsymbol{x}(t; t_0, \boldsymbol{u}) - \boldsymbol{x}(t; t_0, \boldsymbol{v})\| \leq \|\boldsymbol{u} - \boldsymbol{v}\| + L_k \int_{t_0}^{t} \|\boldsymbol{x}(s; t_0, \boldsymbol{u}) - \boldsymbol{x}(s; t_0, \boldsymbol{v})\| ds \tag{2.70}$$

and, invoking Gronwall's lemma again, we obtain

$$\|\boldsymbol{x}(t; t_0, \boldsymbol{u}) - \boldsymbol{x}(t; t_0, \boldsymbol{v})\| \leq e^{L_K(t-t_0)} \|\boldsymbol{u} - \boldsymbol{v}\|, \quad \forall t \in [t_0, T]. \tag{2.71}$$

The last inequality proves the continuity of the mapping $\boldsymbol{v} \mapsto \boldsymbol{x}(t; t_0, \boldsymbol{v})$ on the ball $B(\boldsymbol{x}_0, \eta(T'))$. This completes the proof of Theorem 2.14. □

Let us now consider the special case when system (2.39) is autonomous, that is, the map $\boldsymbol{f}$ is independent of t.

More precisely, we assume that $\boldsymbol{f} : \mathbb{R}^n \to \mathbb{R}^n$ is a locally Lipschitz function. One should think of $\boldsymbol{f}$ as a vector field on $\mathbb{R}^n$.

For any $\boldsymbol{y} \in \mathbb{R}^n$, we set

$$S(t)\boldsymbol{u} := \boldsymbol{x}(t; 0, \boldsymbol{u}),$$

where $\boldsymbol{x}(t; 0, u)$ is the unique saturated solution of the system

$$\boldsymbol{x}' = \boldsymbol{f}(\boldsymbol{x}), \tag{2.72}$$

satisfying the initial condition $\boldsymbol{x}(0) = \boldsymbol{u}$. Theorem 2.14 shows that for any $\boldsymbol{x}_0 \in \mathbb{R}^n$ there exists a $T > 0$ and a neighborhood $U_0 = B(\boldsymbol{x}_0, \eta)$ of $\boldsymbol{x}_0$ such that $S(t)\boldsymbol{u}$ is well defined for any $\boldsymbol{u} \in U_0$ and any $|t| \leq T$. Moreover, the resulting maps

$$U_0 \ni \boldsymbol{u} \mapsto S(t)\boldsymbol{u} \in \mathbb{R}^n$$

are continuous for any $|t| \leq T$. From the local existence and uniqueness theorem, we deduce that the family of maps $S(t) : U_0 \to \mathbb{R}^n$, $-T \leq t \leq T$, has the following properties

$$S(0)\boldsymbol{u} = \boldsymbol{u}, \quad \forall \boldsymbol{u} \in U_0, \tag{2.73}$$

$$\begin{aligned} S(t+s)\boldsymbol{u} = S(t)S(s)\boldsymbol{u}, \ \forall s, t \in [-T, T] \\ \text{such that } |t+s| \leq T, \ S(s)\boldsymbol{u} \in U_0, \end{aligned} \tag{2.74}$$

$$\lim_{t \to 0} S(t)\boldsymbol{u} = \boldsymbol{u}, \quad \forall \boldsymbol{u} \in U_0. \tag{2.75}$$

The family of applications $\{S(t)\}_{|t| \leq T}$ is called the *local flow* or the *continuous local one-parameter group* generated by the vector field $\boldsymbol{f} : \mathbb{R}^n \to \mathbb{R}^n$. From the definition of $S(t)$, we deduce that

$$\boldsymbol{f}(\boldsymbol{u}) = \lim_{t \to 0} \frac{1}{t}\big(S(t)\boldsymbol{u} - \boldsymbol{u} \big), \quad \forall \boldsymbol{u} \in U_0. \tag{2.76}$$

Consider now the differential system

$$\boldsymbol{x}' = \boldsymbol{f}(t, \boldsymbol{x}, \lambda), \quad \lambda \in \Lambda \subset \mathbb{R}^m, \tag{2.77}$$

where $\boldsymbol{f} : \Omega \times \Lambda \to \mathbb{R}^n$ is a continuous function, Ω is an open subset of $\mathbb{R}^{n+1}$, and Λ is an open subset of $\mathbb{R}^m$. Additionally, we will assume that $\boldsymbol{f}$ is locally Lipschitz in $(\boldsymbol{x}, \lambda)$ on $\Omega \times \Lambda$. In other words, for any compact sets $K_1 \subset \Omega$ and $K_2 \subset \Lambda$ there exists a positive constant L such that

$$\begin{aligned} \|\boldsymbol{f}(t, \boldsymbol{x}, \lambda) - \boldsymbol{f}(t, \boldsymbol{y}, \mu)\| \leq L\big(\|\boldsymbol{x} - \boldsymbol{y}\| + \|\lambda - \mu\| \big), \\ \forall (t, \boldsymbol{x}), (t, \boldsymbol{y}) \in K_1, \quad \lambda, \mu \in K_2. \end{aligned} \tag{2.78}$$

Above, we denoted by the same symbol the norms $\| - \|$ on $\mathbb{R}^m$ and $\mathbb{R}^n$.

For any $(t_0, \boldsymbol{x}_0) \in \Omega$, and $\lambda \in \Lambda$, the system (2.77) admits a unique solution $\boldsymbol{x} = \boldsymbol{x}(t; t_0, \boldsymbol{x}_0, \lambda)$ satisfying the initial condition $\boldsymbol{x}(t_0) = \boldsymbol{x}_0$. Loosely speaking, our next result states that the correspondence $\lambda \mapsto \boldsymbol{x}(-; t_0, \boldsymbol{x}_0, \lambda)$ is continuous.

Theorem 2.15 (Continuous dependence on parameters) *Fix a point* $(t_0, \boldsymbol{x}_0, \lambda_0) \in \Omega \times \Lambda$. *Let* $[t_0, T)$ *be the maximal interval of existence on the right of the solution* $\boldsymbol{x}(t; t_0, \boldsymbol{x}_0, \lambda_0)$. *Then, for any* $T' \in [t_0, T)$, *there exists an* $\eta = \eta(T') > 0$ *such that for any* $\lambda \in B(\lambda_0, \eta)$ *the solution* $\boldsymbol{x}(t; t_0, \boldsymbol{x}_0, \lambda)$ *is defined on* $[t_0, T']$. *Moreover, the application*

$$B(\lambda_0, \eta) \ni \lambda \mapsto \boldsymbol{x}(-; t_0, \boldsymbol{x}_0, \lambda) \in C\big([t_0, T'], \mathbb{R}^n\big)$$

is continuous.

Proof The above result is a special case of Theorem 2.14 on the continuous dependence on initial data.

Indeed, if we denote by z the $(n+m)$-dimensional vector $(\boldsymbol{x}, \lambda) \in \mathbb{R}^{n+m}$, and we define

$$\widetilde{\boldsymbol{f}} : \Omega \times \Lambda \to \mathbb{R}^{n+m}, \quad \widetilde{\boldsymbol{f}}(t, \boldsymbol{x}, \lambda) = \big(\boldsymbol{f}(t, \boldsymbol{x}, \lambda), 0\big) \in \mathbb{R}^n \times \mathbb{R}^m,$$

then system (2.77) can be rewritten as

$$z'(t) = \widetilde{\boldsymbol{f}}\big(t, z(t)\big), \tag{2.79}$$

while the initial condition becomes

$$z(t_0) = z_0 := (\boldsymbol{x}_0, \lambda). \tag{2.80}$$

We have thus reduced the problem to investigating the dependence of the solutions $z(t)$ of (2.79) on the initial data. Our assumptions on $\boldsymbol{f}$ show that $\widetilde{\boldsymbol{f}}$ satisfies the assumptions of Theorem 2.14. □

2.7 Differential Inclusions

One of the possible extensions of the concept of a differential equation is to consider instead of the function $\boldsymbol{f} : \Omega \subset \mathbb{R}^{n+1} \to \mathbb{R}^n$ a set-valued, or multi-valued map

$$F : \Omega \to 2^{\mathbb{R}^n},$$

where we recall that for any set S we denote by 2^S the collection of its subsets. In this case, system (2.39) becomes a *differential inclusion*

$$\boldsymbol{x}'(t) \in F\big(t, \boldsymbol{x}(t)\big), \ t \in I, \tag{2.81}$$

to which we associate the initial condition

$$\boldsymbol{x}(t_0) = \boldsymbol{x}_0. \tag{2.82}$$

In general, we cannot expect the existence of a continuously differentiable solution of the Cauchy problem (2.81) and (2.82). Consider, for example, the differential inclusion

$$x' \in \operatorname{Sign} x, \tag{2.83}$$

where Sign : $\mathbb{R} \to 2^{\mathbb{R}^n}$ is given by

$$\operatorname{Sign}(x) = \begin{cases} -1, & x < 0, \\ [-1, 1], & x = 0, \\ 1, & x > 0. \end{cases}$$

Note that, if $x_0 > 0$, then the function

$$x(t) = \begin{cases} t - t_0 + x_0, & t \geq -x_0 + t_0, \\ 0, & t < -x_0 + t_0, \end{cases}$$

is the unique solution of (2.83) on $\mathbb{R} \setminus \{t_0 - x_0\}$. However, it is not a C^1-function since its derivative has a discontinuity at $t_0 - x_0$. Thus, the above function is not a solution in the sense we have adopted so far.

This simple example suggests the need to extend the concept of solution.

Definition 2.1 The function $\boldsymbol{x} : [t_0, T] \to \mathbb{R}^n$ is called a *Carathéodory solution* of the differential inclusion (2.81) if the following hold.

(i) The function $\boldsymbol{x}(t)$ is absolutely continuous on $[t_0, T]$.
(ii) There exists a negligible set $N \subset [t_0, T]$ such that, for any $t \in [t_0, T] \setminus N$, the function $x(t)$ is differentiable at t and $\boldsymbol{x}'(t) \in F(t, \boldsymbol{x}(t))$.

According to Lebesgue's theorem (see e.g. [12, Sect. 33]), an absolutely continuous function $\boldsymbol{x} : [t_0, T] \to \mathbb{R}^n$ is almost everywhere differentiable on the interval $[t_0, T]$.

Differential inclusions naturally appear in the modern theory of variational calculus and of control systems. An important source of differential inclusion is represented by differential equations with a discontinuous right-hand side. More precisely, if $f = f(t, x)$ is discontinuous in x, then the Cauchy problem (2.1) does not have a Carathéodory solution, but this might happen if we extend f to a multi-valued mapping $(t, x) \to F(t, x)$. (This happens for Eq. (2.83), where the discontinuous function $\frac{x}{|x|}$ was extended to Sign x.) In this section, we will investigate a special class of differential inclusions known as *evolution variational inequalities*. They were introduced in mathematics, in a more general context, by *G. Stampacchia* (1922–1978) and *J.L. Lions* (1928–2001). To state and solve such problems, we need to make a brief digression into (finite-dimensional) convex analysis.

Recall that a subset $C \subset \mathbb{R}^n$ is *convex* if

$$(1-t)\boldsymbol{x} + t\boldsymbol{y} \in C, \quad \forall \boldsymbol{x}, \boldsymbol{y} \in C, \quad \forall t \in [0,1].$$

Geometrically, this means that for any two points in C the line segment connecting them is entirely contained in C. Given a closed convex set $C \subset \mathbb{R}^n$ and $\boldsymbol{x}_0 \in C$, we set

$$N_C(\boldsymbol{x}_0) := \big\{ \boldsymbol{w} \in \mathbb{R}^n; \;\; (\boldsymbol{w}, \boldsymbol{y} - \boldsymbol{x}_0) \leq 0, \;\; \forall \boldsymbol{y} \in C \big\}. \tag{2.84}$$

The set $N_C(\boldsymbol{x}_0)$ is a closed convex cone called the (outer) *normal cone* of C at the point $\boldsymbol{x}_0 \in C$. We extend N_C to a multi-valued map $N_C : \mathbb{R}^n \to 2^{\mathbb{R}^n}$ by setting

$$N_C(\boldsymbol{x}) = \emptyset, \quad \forall \boldsymbol{x} \in \mathbb{R}^n \setminus C.$$

Example 2.5 (a) If C is a convex domain in $\mathbb{R}^n$ with smooth boundary and $\boldsymbol{x}_0$ is a point on the boundary, then $N_C(\boldsymbol{x}_0)$ is the cone spanned by the unit outer normal to ∂C at $\boldsymbol{x}_0$. If $\boldsymbol{x}_0$ is in the interior of C, then $N_C(\boldsymbol{x}_0) = \{0\}$.
(b) If $C \subset \mathbb{R}^n$ is a vector subspace, then for any $\boldsymbol{x} \in C$ we have $N_C(\boldsymbol{x}) = C^{\perp}$, the orthogonal complement of C in $\mathbb{R}^n$.

To any closed convex set $C \subset \mathbb{R}^n$ there corresponds a projection

$$P_C : \mathbb{R}^n \to C$$

that associates to each $\boldsymbol{x} \in \mathbb{R}^n$ the point in C closest to $\boldsymbol{x}$ with respect to the Euclidean distance. The next result makes this precise.

Lemma 2.1 *Let C be a closed convex subset of $\mathbb{R}^n$. Then the following hold.*
(a) *For any $\boldsymbol{x} \in \mathbb{R}^n$ there exists a unique point $\boldsymbol{y} \in C$ such that*

$$\|\boldsymbol{x} - \boldsymbol{y}\|_{\mathbf{e}} = \operatorname{dist}(\boldsymbol{x}, C) := \inf_{z \in C} \|\boldsymbol{x} - z\|_{\mathbf{e}}. \tag{2.85}$$

We denote by $P_C(\boldsymbol{x})$ this unique point in C, and we will refer to the resulting map $P_C : \mathbb{R}^n \to C$ as the orthogonal projection onto C.
(b) *The map $P_C : \mathbb{R}^n \to C$ satisfies the following properties:*

$$\begin{aligned} \boldsymbol{x} - P_C\boldsymbol{x} \in N_C\big(P_C\boldsymbol{x}\big), \;\; &\textit{that is,} \\ \big(\boldsymbol{x} - P_C\boldsymbol{x}, \boldsymbol{y} - P_C\boldsymbol{x}\big) \leq 0, \;\; \forall \boldsymbol{x} \in \mathbb{R}^n, \;\; &\boldsymbol{y} \in C. \end{aligned} \tag{2.86}$$

$$\|P_C\boldsymbol{x} - P_C\boldsymbol{z}\|_{\mathbf{e}} \leq \|\boldsymbol{x} - \boldsymbol{y}\|_{\mathbf{e}}, \quad \forall \boldsymbol{x}, \boldsymbol{z} \in \mathbb{R}^n. \tag{2.87}$$

Proof (a) There exists a sequence $(\boldsymbol{y}_\nu)$ in C such that

$$\operatorname{dist}(\boldsymbol{x}, C) \leq \|\boldsymbol{x} - \boldsymbol{y}_\nu\|_{\mathbf{e}} \leq \operatorname{dist}(\boldsymbol{x}, C) + \frac{1}{\nu}. \tag{2.88}$$

The sequence $(\boldsymbol{y}_\nu)$ is obviously bounded and thus it has a convergent subsequence $(\boldsymbol{y}_{\nu_k})$. Its limit $\boldsymbol{y}$ is a point in C since C is closed. Moreover, inequalities (2.88) imply that

$$\|\boldsymbol{x}-\boldsymbol{y}\|_{\mathbf{e}} = \operatorname{dist}(\boldsymbol{x}, C).$$

This completes the proof of the existence part of (a).

Let us prove the uniqueness statement. Let $\boldsymbol{y}_1, \boldsymbol{y}_2 \in C$ such that

$$\|\boldsymbol{x}-\boldsymbol{y}_1\|_{\mathbf{e}} = \|\boldsymbol{x}-\boldsymbol{y}_2\|_{\mathbf{e}} = \operatorname{dist}(\boldsymbol{x}, C).$$

Since C is convex, we deduce that

$$y_0 := \frac{1}{2}(y_1 + y_2) \in C.$$

From the triangle inequality we deduce that

$$\begin{aligned}\operatorname{dist}(\boldsymbol{x}, C) &\leq \|\boldsymbol{x}-\boldsymbol{y}_0\|_{\mathbf{e}} \\ &\leq \frac{1}{2}\Big(\|\boldsymbol{x}-\boldsymbol{y}_1\|_{\mathbf{e}} + \|\boldsymbol{x}-\boldsymbol{y}_2\|_{\mathbf{e}}\Big) \\ &= \operatorname{dist}(\boldsymbol{x}, C).\end{aligned}$$

Hence

$$\|\boldsymbol{x}-\boldsymbol{y}_0\|_{\mathbf{e}} = \|\boldsymbol{x}-\boldsymbol{y}_1\|_{\mathbf{e}} = \|\boldsymbol{x}-\boldsymbol{y}_2\|_{\mathbf{e}} = \operatorname{dist}(\boldsymbol{x}, C). \tag{2.89}$$

One the other hand, we have the parallelogram identity

$$\left\|\frac{1}{2}(y+z)\right\|_{\mathbf{e}}^2 + \left\|\frac{1}{2}(y-z)\right\|_{\mathbf{e}}^2 = \frac{1}{2}\Big(\|y\|_{\mathbf{e}}^2 + \|z\|_{\mathbf{e}}^2\Big), \quad \forall \boldsymbol{y}, z \in \mathbb{R}^n.$$

If in the above equality we let $\boldsymbol{y} = \boldsymbol{x} - \boldsymbol{y}_1, z = \boldsymbol{x} - \boldsymbol{y}_2$, then we conclude from (2.89) that

$$\|\boldsymbol{y}_1 - \boldsymbol{y}_2\|_{\mathbf{e}}^2 = 0.$$

This completes the proof of the uniqueness.

(b) To prove (2.86), we start with the defining inequality

$$\|\boldsymbol{x} - P_C\boldsymbol{x}\|_{\mathbf{c}}^2 \leq \|\boldsymbol{x}-\boldsymbol{y}\|^2, \quad \forall \boldsymbol{y} \in C.$$

Consider now the function

$$f_{\boldsymbol{y}} : [0,1] \to \mathbb{R}, \quad f_{\boldsymbol{y}}(t) = \|\boldsymbol{x}-\boldsymbol{y}_t\|^2 - \|\boldsymbol{x} - P_C\boldsymbol{x}\|_{\mathbf{e}}^2,$$

where

$$y_t = (1-t)P_C\boldsymbol{x} + t\boldsymbol{y} = P_C(x) + t(\boldsymbol{y} - P_C\boldsymbol{x}).$$

We have $f_{\boldsymbol{y}}(t) \geq 0$, $\forall t \geq 0$ and $f_{\boldsymbol{y}}(0) = 0$. Thus

$$f'_{\boldsymbol{y}}(0) \geq 0.$$

Observing that

$$\begin{aligned} f_{\boldsymbol{y}}(t) &= \|(\boldsymbol{x} - P_C\boldsymbol{x}) - t(\boldsymbol{y} - P_C\boldsymbol{x})\|_{\mathbf{e}}^2 - \|\boldsymbol{x} - P_C\boldsymbol{x}\|_{\mathbf{e}}^2 \\ &= t^2\|\boldsymbol{y} - P_C\boldsymbol{x}\|_{\mathbf{e}}^2 - 2t\big(\boldsymbol{x} - P_C\boldsymbol{x},\ \boldsymbol{y} - P_C\boldsymbol{x}\big), \end{aligned}$$

we deduce that

$$f'_{\boldsymbol{y}}(0) = -2\big(\boldsymbol{x} - P_C\boldsymbol{x},\ \boldsymbol{y} - P_C\boldsymbol{x}\big) \geq 0,\quad \forall \boldsymbol{y} \in C.$$

This proves (2.86).

To prove (2.87), let $z \in \mathbb{R}^n$ and set

$$\boldsymbol{u} := \boldsymbol{x} - P_C\boldsymbol{x},\quad \boldsymbol{v} := z - P_C z,\quad w := P_C z - P_C\boldsymbol{x}.$$

From (2.86) we deduce that

$$\boldsymbol{u} \in N_C(P_C\boldsymbol{x}),\quad \boldsymbol{v} \in N_C(P_C z),$$

so that $(\boldsymbol{u}, \boldsymbol{w}) \leq 0 \leq (\boldsymbol{v}, \boldsymbol{w})$ and thus

$$(\boldsymbol{w}, \boldsymbol{u} - \boldsymbol{w}) \leq 0. \tag{2.90}$$

On the other hand, we have $\boldsymbol{x} - z = \boldsymbol{u} - \boldsymbol{w} - \boldsymbol{v}$, so that

$$\begin{aligned} \|\boldsymbol{x} - z\|_{\mathbf{e}}^2 &= \|(\boldsymbol{u} - \boldsymbol{v}) - \boldsymbol{w}\|^2 \\ &= \|\boldsymbol{w}\|_{\mathbf{e}}^2 + \|\boldsymbol{u} - \boldsymbol{v}\|_{\mathbf{e}}^2 - 2(\boldsymbol{w}, \boldsymbol{u} - \boldsymbol{w}) \\ &\overset{(2.90)}{\geq} \|\boldsymbol{w}\|_{\mathbf{e}}^2 \\ &= \|P_C\boldsymbol{x} - P_C z\|_{\mathbf{e}}^2. \end{aligned}$$

□

Suppose that K is a closed convex subset of $\mathbb{R}^n$. Fix real numbers $t_0 < T$, a continuous function $\boldsymbol{g} : [t_0, T] \to \mathbb{R}^n$ and a (globally) Lipschitz map $\boldsymbol{f} : \mathbb{R}^n \to \mathbb{R}^n$. We want to investigate the differential inclusion

$$\begin{aligned}\boldsymbol{x}'(t) &\in \boldsymbol{f}\big(\boldsymbol{x}(t)\big) + \boldsymbol{g}(t) - N_K\big(\boldsymbol{x}(t)\big), \ \text{a.e. } t \in (t_0, T)\\ \boldsymbol{x}(t_0) &= \boldsymbol{x}_0.\end{aligned} \tag{2.91}$$

This differential inclusion can be rewritten as an evolution variational inequality

$$\boldsymbol{x}(t) \in K, \quad \forall t \in [0, T], \tag{2.92}$$

$$\big(\boldsymbol{x}'(t) - \boldsymbol{f}\big(\boldsymbol{x}(t)\big) - \boldsymbol{g}(t), \boldsymbol{y} - \boldsymbol{x}(t)\big) \geq 0, \ \text{a.e. } t \in (t_0, T), \ \forall \boldsymbol{y} \in K, \tag{2.93}$$

$$\boldsymbol{x}(t_0) = \boldsymbol{x}_0. \tag{2.94}$$

Theorem 2.16 *Suppose that $\boldsymbol{x}_0 \in K$ and $g : [t_0, T] \to \mathbb{R}^n$ is a continuously differentiable function. Then the initial value problem* (2.91) *admits a unique Carathéodory solution $\boldsymbol{x} : [t_0, T] \to \mathbb{R}^n$. Moreover,*

$$\boldsymbol{x}'(t) = \boldsymbol{f}(\boldsymbol{x}(t)) + \boldsymbol{g}(t) - P_{N_K(\boldsymbol{x}(t))}(\boldsymbol{f}(\boldsymbol{x}(t)) + \boldsymbol{g}(t)), \ a.e. \ t \in (t_0, T). \tag{2.95}$$

Proof For simplicity, we denote by P the orthogonal projection P_K onto K defined in Lemma 2.1. Define the map

$$\Gamma : \mathbb{R}^n \to \mathbb{R}^n, \quad \Gamma\boldsymbol{x} = P\boldsymbol{x} - \boldsymbol{x}.$$

Note that $-\Gamma\boldsymbol{x} \in N_K(P\boldsymbol{x})$ and $\|\Gamma\boldsymbol{x}\|_{\mathbf{e}} = \text{dist}(\boldsymbol{x}, K)$. Moreover, Γ is dissipative, that is,

$$\big(\Gamma\boldsymbol{x} - \Gamma\boldsymbol{y}, \boldsymbol{x} - \boldsymbol{y}\big) \leq 0, \quad \forall \boldsymbol{x}, \boldsymbol{y} \in \mathbb{R}^n. \tag{2.96}$$

Indeed,

$$\begin{aligned}\big(\Gamma\boldsymbol{x} - \Gamma\boldsymbol{y}, \boldsymbol{x} - \boldsymbol{y}\big) &= \big(P\boldsymbol{x} - P\boldsymbol{y}, \boldsymbol{x} - \boldsymbol{y}\big) - \|\boldsymbol{x} - \boldsymbol{y}\|_{\mathbf{e}}^2\\ &\leq \|P\boldsymbol{x} - P\boldsymbol{y}\|_{\mathbf{e}} \cdot \|\boldsymbol{x} - \boldsymbol{y}\|_{\mathbf{e}} - \|\boldsymbol{x} - \boldsymbol{y}\|_{\mathbf{e}}^2 \overset{(2.87)}{\leq} 0.\end{aligned}$$

We will obtain the solution of (2.91) as the limit of the solutions $\{\boldsymbol{x}_\varepsilon\}_{\varepsilon>0}$ of the approximative Cauchy problem

$$\begin{aligned}\boldsymbol{x}_\varepsilon'(t) &= \boldsymbol{f}\big(\boldsymbol{x}_\varepsilon(t)\big) + \boldsymbol{g}(t) + \frac{1}{\varepsilon}\Gamma\boldsymbol{x}_\varepsilon(t),\\ \boldsymbol{x}_\varepsilon(t_0) &= \boldsymbol{x}_0.\end{aligned} \tag{2.97}$$

For any $\varepsilon > 0$, the map $F_\varepsilon : \mathbb{R}^n \to \mathbb{R}^n$, $F_\varepsilon(\boldsymbol{x}) = \boldsymbol{f}(\boldsymbol{x}) + \frac{1}{\varepsilon}\Gamma\boldsymbol{x}$ is Lipschitz. Hence, the Cauchy problem (2.97) has a unique right-saturated solution $\boldsymbol{x}_\varepsilon(t)$ defined on an interval $[t_0, T_\varepsilon)$. Since F_ε is globally Lipschitz, it follows that $\boldsymbol{x}_\varepsilon$ is defined over $[t_0, T]$. (See Example 2.1.)

Taking the scalar product of (2.97) with $\boldsymbol{x}_\varepsilon(t) - \boldsymbol{x}_0$, and observing that $\Gamma\boldsymbol{x}_0 = 0$, $\boldsymbol{x}_0 \in K$, we deduce from (2.96) that

$$\begin{aligned}\frac{1}{2}\frac{d}{dt}\|\boldsymbol{x}_\varepsilon(t)-\boldsymbol{x}_0\|_{\mathbf{e}}^2 &\leq \big(\boldsymbol{f}(\boldsymbol{x}_\varepsilon(t))+\boldsymbol{g}(t),\ \boldsymbol{x}_\varepsilon(t)-\boldsymbol{x}_0\big)\\ &\leq \|\boldsymbol{f}(\boldsymbol{x}_\varepsilon(t))-\boldsymbol{f}(\boldsymbol{x}_0)\|_{\mathbf{e}}\cdot\|\boldsymbol{x}_\varepsilon-\boldsymbol{x}_0\|_{\mathbf{e}}\\ &\quad+\big(\|\boldsymbol{f}(\boldsymbol{x}_0)\|_{\mathbf{e}}+\|\boldsymbol{g}(t)\|_{\mathbf{e}}\big)\cdot\|\boldsymbol{x}_\varepsilon-\boldsymbol{x}_0\|_{\mathbf{e}}.\end{aligned}$$

Hence, if L denotes the Lipschitz constant of $\boldsymbol{f}$, we have

$$\begin{aligned}\frac{1}{2}\frac{d}{dt}\|\boldsymbol{x}_\varepsilon(t)-\boldsymbol{x}_0\|_{\mathbf{e}}^2 &\leq L\|\boldsymbol{x}_\varepsilon(t)-\boldsymbol{x}_0\|_{\mathbf{e}}^2+\frac{1}{2}\big(\|\boldsymbol{f}(\boldsymbol{x}_0)\|_{\mathbf{e}}+\|\boldsymbol{g}(t)\|_{\mathbf{e}}\big)^2\\ &\quad+\frac{1}{2}\|\boldsymbol{x}_\varepsilon(t)-\boldsymbol{x}_0\|_{\mathbf{e}}^2.\end{aligned}$$

We set

$$M:=\sup_{t\in[t_0,T]}\Big(\|\boldsymbol{f}(\boldsymbol{x}_0)\|_{\mathbf{e}}+\|\boldsymbol{g}(t)\|_{\mathbf{e}}\Big)^2,\ L'=2L+1,$$

and we get

$$\frac{d}{dt}\|\boldsymbol{x}_\varepsilon(t)-\boldsymbol{x}_0\|_{\mathbf{e}}^2\leq L'\|\boldsymbol{x}_\varepsilon(t)-\boldsymbol{x}_0\|_{\mathbf{e}}^2+M.$$

Gronwall's lemma now yields the following ε*-independent* upper bound

$$\|\boldsymbol{x}_\varepsilon(t)-\boldsymbol{x}_0\|_{\mathbf{e}}^2\leq Me^{L'(t-t_0)},\quad \forall t\in[t_0,T]. \tag{2.98}$$

Using (2.97), we deduce that

$$\begin{aligned}\frac{d}{dt}\big(\boldsymbol{x}_\varepsilon(t+h)-\boldsymbol{x}_\varepsilon(t)\big) &= \boldsymbol{f}(\boldsymbol{x}_\varepsilon(t+h))-\boldsymbol{f}(\boldsymbol{x}_\varepsilon(t))+\boldsymbol{g}(t+h)-\boldsymbol{g}(t)\\ &\quad+\frac{1}{\varepsilon}\big(\Gamma\boldsymbol{x}_\varepsilon(t+h)-\Gamma\boldsymbol{x}_\varepsilon(t)\big).\end{aligned}$$

Let $h>0$. Taking the scalar product with $\boldsymbol{x}_\varepsilon(t+h)-\boldsymbol{x}_\varepsilon(t)$ of both sides of the above equality and using the dissipativity of Γ, we deduce that for any $t\in[t_0,T]$ we have

$$\begin{aligned}\frac{1}{2}\frac{d}{dt}\|\boldsymbol{x}_\varepsilon(t+h)-\boldsymbol{x}_\varepsilon(t)\|_{\mathbf{e}}^2 &\leq \big(\boldsymbol{f}(\boldsymbol{x}_\varepsilon(t+h))-\boldsymbol{f}(\boldsymbol{x}_\varepsilon(t)),\boldsymbol{x}_\varepsilon(t+h)-\boldsymbol{x}_\varepsilon(t)\big)\\ &\quad+\big(\boldsymbol{g}(t+h)-\boldsymbol{g}(t),\boldsymbol{x}_\varepsilon(t+h)-\boldsymbol{x}_\varepsilon(t)\big)\\ \leq L\|\boldsymbol{x}_\varepsilon(t+h)-\boldsymbol{x}_\varepsilon(t)\|_{\mathbf{e}}^2 &+\frac{1}{2}\Big(\|\boldsymbol{g}(t+h)-\boldsymbol{g}(t)\|_{\mathbf{e}}^2+\|\boldsymbol{x}_\varepsilon(t+h)-\boldsymbol{x}_\varepsilon(t)\|_{\mathbf{e}}^2\Big),\end{aligned}$$

so that, setting again $L'=2L+1$, we obtain by integration

$$\|\boldsymbol{x}_\varepsilon(t+h)-\boldsymbol{x}_\varepsilon(t)\|_{\mathbf{e}}^2 \le \|\boldsymbol{x}_\varepsilon(h)-\boldsymbol{x}_0\|_{\mathbf{e}}^2 + \int_{t_0}^t \|\boldsymbol{g}(s+h)-\boldsymbol{g}(s)\|_{\mathbf{e}}^2 ds$$
$$+L'\int_{t_0}^t \|\boldsymbol{x}_\varepsilon(s+h)-\boldsymbol{x}_\varepsilon(s)\|_{\mathbf{e}}^2 ds,$$

for all $t \in [t_0, T-h]$. Since g is a C^1-function, there exists a $C_0 > 0$ such that

$$\|\boldsymbol{g}(s+h)-\boldsymbol{g}(s)\|_{\mathbf{e}} \le C_0 h, \quad \forall t \in [t_0, T-h].$$

Hence, $\forall t \in [t_0, T-h]$, we have

$$\|\boldsymbol{x}_\varepsilon(t+h)-\boldsymbol{x}_\varepsilon(t)\|_{\mathbf{e}}^2 \le \|\boldsymbol{x}_\varepsilon(h)-\boldsymbol{x}_0\|_{\mathbf{e}} + C_0^2h^2(T-t_0)$$
$$+ L'\int_{t_0}^t \|\boldsymbol{x}_\varepsilon(s+h)-\boldsymbol{x}_\varepsilon(s)\|_{\mathbf{e}}^2 ds.$$

Using Gronwall's lemma once again, we deduce that

$$\|\boldsymbol{x}_\varepsilon(t+h)-\boldsymbol{x}_\varepsilon(t)\|_{\mathbf{e}}^2 \le \Big(\|\boldsymbol{x}_\varepsilon(h)-\boldsymbol{x}_0\|_{\mathbf{e}} + C_0^2h^2(T-t_0)\Big)e^{L'(t-t_0)}$$
$$\stackrel{(2.98)}{\le} \Big(Me^{L'h} + C_0^2h^2(T-t_0)\Big)e^{L'(t-t_0)}.$$

Thus, for some constant $C_1 > 0$, *independent of* ε *and* h, we have

$$\|\boldsymbol{x}_\varepsilon(t+h)-\boldsymbol{x}_\varepsilon(t)\|_{\mathbf{e}} \le C_1 h, \quad \forall t \in [t_0, T-h].$$

Thus

$$\|\boldsymbol{x}_\varepsilon'(t)\| \le C_1, \quad \forall t \in [0, T]. \tag{2.99}$$

From the equality

$$\boldsymbol{x}_\varepsilon(t)-\boldsymbol{x}_\varepsilon(s) = \int_s^t \boldsymbol{x}_\varepsilon'(\tau)d\tau$$

we find that

$$\|\boldsymbol{x}_\varepsilon(t)-\boldsymbol{x}_\varepsilon(s)\|_{\mathbf{e}} \le C_1|t-s|, \quad \forall t, s \in [t_0, T]. \tag{2.100}$$

This shows that the family $\{\boldsymbol{x}_\varepsilon\}_{\varepsilon>0}$ is uniformly bounded and equicontinuous on $[t_0, T]$. Arzelà's theorem now implies that there exists a subsequence (for simplicity, denoted by ε) and a continuous function $\boldsymbol{x} : [t_0, T] \to \mathbb{R}^n$ such that $\boldsymbol{x}_\varepsilon(t)$ converges uniformly to $\boldsymbol{x}(t)$ on $[t_0, T]$ as $\varepsilon \to 0$.

Passing to the limit in (2.100), we deduce that the limit function $\boldsymbol{x}(t)$ is Lipschitz on $[t_0, T]$. In particular, $\boldsymbol{x}(t)$ is absolutely continuous and almost everywhere differentiable on this interval. From (2.97) and (2.99), it follows that there exists a constant $C_2 > 0$, *independent of* ε, such that

$$\operatorname{dist}(\boldsymbol{x}_\varepsilon(t), K) = \|\Gamma \boldsymbol{x}_\varepsilon(t)\|_{\mathbf{e}} \leq C_2\varepsilon, \quad \forall t \in [t_0, T].$$

This proves that $\operatorname{dist}(\boldsymbol{x}(t), K) = 0$, $\forall t$, that is, $\boldsymbol{x}(t) \in K$, $\forall t \in [t_0, T]$.

We can now prove inequality (2.93). To do this, we fix a point t where the function $\boldsymbol{x}$ is differentiable (we saw that this happens for almost any $t \in [t_0, T]$). From (2.97) and (2.86), we deduce that for almost all $s \in [t_0, T]$ and any $z \in K$ we have

$$\frac{1}{2}\frac{d}{dt}\|\boldsymbol{x}_\varepsilon(s) - z\|_{\mathbf{e}}^2 \leq \big(\boldsymbol{f}(\boldsymbol{x}_\varepsilon(s)) + \boldsymbol{g}(s),\ \boldsymbol{x}_\varepsilon(s) - z\big).$$

Integrating from t to $t + h$, we deduce that

$$\begin{aligned}\frac{1}{2}(\|\boldsymbol{x}_\varepsilon(t+h) - z\|_{\mathbf{e}}^2 - \|\boldsymbol{x}_\varepsilon(t) - z\|_{\mathbf{e}}^2)&\\ \leq \int_t^{t+h} (\boldsymbol{f}(\boldsymbol{x}_\varepsilon(s)) + \boldsymbol{g}(s), \boldsymbol{x}_\varepsilon(s) - z)ds,\ \forall z \in K.&\end{aligned}$$

Now, let us observe that, for any $\boldsymbol{u}, \boldsymbol{v} \in \mathbb{R}^n$, we have

$$\frac{1}{2}\big(\|\boldsymbol{u} + \boldsymbol{v}\|_{\mathbf{e}}^2 - \|\boldsymbol{v}\|_{\mathbf{e}}^2\big) \geq (\boldsymbol{u}, \boldsymbol{v}).$$

Using this inequality with $\boldsymbol{u} = \boldsymbol{x}_\varepsilon(t+h) - \boldsymbol{x}_\varepsilon(t)$, $\quad \boldsymbol{v} = \boldsymbol{x}_\varepsilon(t) - z$, we get

$$\begin{aligned}\frac{1}{h}\big(\boldsymbol{x}_\varepsilon(t+h) - \boldsymbol{x}_\varepsilon(t),\ \boldsymbol{x}_\varepsilon(t) - z\big) &\leq \frac{1}{2h}\big(\|\boldsymbol{x}_\varepsilon(t+h) - z\|_{\mathbf{e}}^2 - \|\boldsymbol{x}_\varepsilon(t) - z\|_{\mathbf{e}}^2\big)\\ &\leq \frac{1}{h}\int_t^{t+h}\big(\boldsymbol{f}(\boldsymbol{x}_\varepsilon(s)) + \boldsymbol{g}(s),\ \boldsymbol{x}_\varepsilon(s) - z\big)ds, \quad \forall z \in K.\end{aligned}$$

Letting $\varepsilon \to 0$, we find

$$\frac{1}{h}\big(\boldsymbol{x}(t+h) - \boldsymbol{x}(t),\ \boldsymbol{x}(t) - z\big) \leq \frac{1}{h}\int_t^{t+h}\big(\boldsymbol{f}(\boldsymbol{x}(s)) + \boldsymbol{g}(s),\ \boldsymbol{x}(s) - z\big)ds, \quad \forall z \in K.$$

Finally, letting $h \to 0$, we obtain

$$\big(\boldsymbol{x}'(t) - \boldsymbol{f}(\boldsymbol{x}(t)) - \boldsymbol{g}(t), \boldsymbol{x}(t) - z) \leq 0, \quad \forall z \in K.$$

This is precisely (2.93).

The uniqueness of the solution now follows easily. Suppose that $\boldsymbol{x}, \boldsymbol{y}$ are solutions of (2.69)–(2.94). We obtain from (2.93) that

$$\big(\boldsymbol{x}'(t) - \boldsymbol{f}(\boldsymbol{x}(t)) - \boldsymbol{g}(t), \boldsymbol{x}(t) - \boldsymbol{y}(t)) \leq 0,$$

$$\big(\boldsymbol{y}'(t) - \boldsymbol{f}(\boldsymbol{y}(t)) - \boldsymbol{g}(t), \boldsymbol{y}(t) - \boldsymbol{x}(t)) \leq 0,$$

so that

$$\big(\boldsymbol{x}'(t) - \boldsymbol{y}'(t) - (\,\boldsymbol{f}(\boldsymbol{x}(t)) - \boldsymbol{f}(\boldsymbol{y}(t))\,), \boldsymbol{x}(t) - \boldsymbol{y}(t)\big) \leq 0$$

which finally implies

$$\frac{1}{2}\frac{d}{dt}\|\boldsymbol{x}(t) - \boldsymbol{y}(t)\|_{\mathbf{e}}^2 \leq L\|\boldsymbol{x}(t) - \boldsymbol{y}(t)\|_{\mathbf{e}}^2,$$

for almost all $t \in [t_0, T]$. Integrating and using the fact that $t \mapsto \|\boldsymbol{x}(t) - \boldsymbol{y}(t)\|_{\mathbf{e}}^2$ is Lipschitz, we deduce that

$$\|\boldsymbol{x}(t) - \boldsymbol{y}(t)\|_{\mathbf{e}}^2 \leq 2L \int_{t_0}^{t} \|\boldsymbol{x}(s) - \boldsymbol{y}(s)\|_{\mathbf{e}}^2 ds.$$

Gronwall's lemma now implies $\boldsymbol{x}(t) = \boldsymbol{y}(t), \forall t$. We have one last thing left to prove, namely, (2.95). Let us observe that (2.91) implies

$$\frac{d}{ds}\boldsymbol{x}(t+s) - \boldsymbol{f}(\boldsymbol{x}(t+s)) - \boldsymbol{g}(t+s) \in -N_K(\boldsymbol{x}(t+s))$$

for almost all t, s. On the other hand, using (2.84), we deduce that

$$\big(\boldsymbol{u} - \boldsymbol{v}, \boldsymbol{x}(t+s) - \boldsymbol{x}(s)\big) \geq 0, \quad \forall \boldsymbol{u} \in N_K(\boldsymbol{x}(t+s)), \quad \boldsymbol{v} \in N_K(\boldsymbol{x}(t)).$$

Hence

$$\begin{aligned}\frac{1}{2}\frac{d}{ds}\|\boldsymbol{x}(t+s) - \boldsymbol{x}(t)\|_{\mathbf{e}}^2 &= \Big(\frac{d}{ds}\boldsymbol{x}(t+s),\ \boldsymbol{x}(t+s) - \boldsymbol{x}(t)\Big)\\ &\leq \big(-\boldsymbol{v} + \boldsymbol{f}(\boldsymbol{x}(t+s)) + \boldsymbol{g}(t+s),\ \boldsymbol{x}(t+s) - \boldsymbol{x}(t)\big),\end{aligned}$$

$\forall \boldsymbol{v} \in N_K(\boldsymbol{x}(t))$. Integrating with respect to s on $[0, h]$, we deduce that

$$\begin{aligned}\frac{1}{2}\|\boldsymbol{x}(t+h) - \boldsymbol{x}(t)\|_{\mathbf{e}}^2 &\leq \int_0^h (-\boldsymbol{v} + \boldsymbol{f}(\boldsymbol{x}(t+s)) + \boldsymbol{g}(t+s), \boldsymbol{x}(t+s) - \boldsymbol{x}(t))ds\\ &\leq \int_0^h \|-\boldsymbol{v} + \boldsymbol{f}(\boldsymbol{x}(t+s)) + \boldsymbol{g}(t+s)\|_{\mathbf{e}} \cdot \|\boldsymbol{x}(t+s) - \boldsymbol{x}(t)\|_{\mathbf{e}}\, ds.\end{aligned}$$

Using Proposition 1.2, we conclude that

$$\|\boldsymbol{x}(t+h) - \boldsymbol{x}(t)\|_{\mathbf{e}} \leq \int_0^h \|-\boldsymbol{v} + \boldsymbol{f}(\boldsymbol{x}(t+s)) + \boldsymbol{g}(t+s)\|_{\mathbf{e}} ds, \quad \forall h, \quad \forall \boldsymbol{v} \in N_K(\boldsymbol{x}(t)).$$

Dividing by $h > 0$ and letting $h \to 0$, we deduce that

$$\|\boldsymbol{x}'(t)\|_{\mathbf{e}} \leq \|-\boldsymbol{v} + \boldsymbol{f}(\boldsymbol{x}(t)) + \boldsymbol{g}(t)\|_{\mathbf{e}}, \quad \forall \boldsymbol{v} \in N_K(\boldsymbol{x}(t)).$$

This means that $\boldsymbol{f}(\boldsymbol{x}(t)) + \boldsymbol{g}(t) - \boldsymbol{x}'(t)$ is the point in $N_K(\boldsymbol{x}(t))$ closest to $\boldsymbol{f}(\boldsymbol{x}(t)) + \boldsymbol{g}(t)$. This is precisely the statement (2.95). □

Remark 2.7 If the solution $\varphi(t)$ of the Cauchy problem

$$\varphi' = \boldsymbol{f}(\varphi) + \boldsymbol{g}, \quad \forall t \in [t_0, T], \quad \varphi(t_0) = \boldsymbol{x}_0, \tag{2.101}$$

stays in K for all $t \in [t_0, T]$, then φ coincides with the unique solution $\boldsymbol{x}(t)$ of (2.92)–(2.94).

Indeed, if we subtract (2.101) from (2.91) and we take the scalar product with $\boldsymbol{x}(t) - \varphi(t)$, then we obtain the inequality

$$\frac{1}{2}\|\boldsymbol{x}(t) - \varphi(t)\|_{\mathbf{e}}^2 \leq L\|\boldsymbol{x}(t) - \varphi(t)\|_{\mathbf{e}}^2, \quad \forall t \in [t_0, T].$$

Gronwall's lemma now implies $\boldsymbol{x} \equiv \varphi$.

Example 2.6 Consider a particle of unit mass that is moving in a planar domain $K \subset \mathbb{R}^2$ under the influence of a homogeneous force field $\boldsymbol{F}(t)$. We assume that K is convex; see Fig. 2.4.

If we denote by $\boldsymbol{g}(t)$ an antiderivative of $\boldsymbol{F}(t)$, and by $\boldsymbol{x}(t)$ the position of the particle at time t, then, intuitively, the motion ought to be governed by the differential equations

$$\begin{aligned} \boldsymbol{x}'(t) &= \boldsymbol{g}(t), \ \text{ if } \boldsymbol{x}(t) \in \operatorname{int} K \\ \boldsymbol{x}'(t) &= \boldsymbol{g}(t) - P_{N(\boldsymbol{x}(t))}\boldsymbol{g}(t), \ \text{ if } \boldsymbol{x}(t) \in \partial K, \end{aligned} \tag{2.102}$$

where $N(\boldsymbol{x}(t))$ is the half-line starting at $\boldsymbol{x}(t)$, normal to ∂K and pointing towards the exterior of K; see Fig. 2.4. (If ∂K is not smooth, then $N_K(\boldsymbol{x}(t))$ is a cone pointed at $\boldsymbol{x}(t)$.) Thus $\boldsymbol{x}(t)$ is the solution of the evolution variational inequation

$$\begin{aligned} \boldsymbol{x}(t) &\in K, \ \forall t \geq 0, \\ \boldsymbol{x}'(t) &\in \boldsymbol{g}(t) - N(\boldsymbol{x}(t)), \ \forall t > 0. \end{aligned}$$

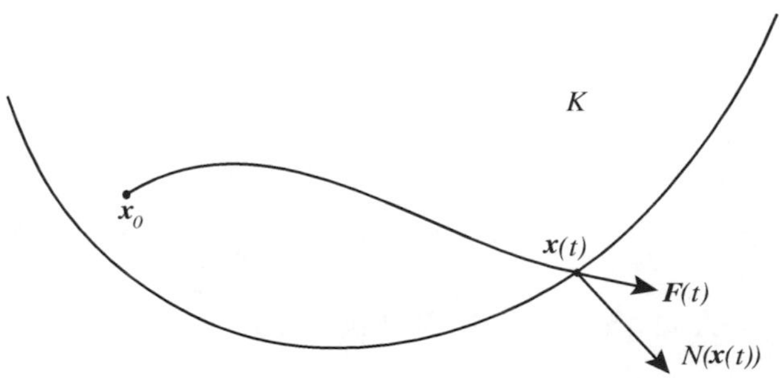

Fig. 2.4 Motion of a particle confined to a convex region

Theorem 2.16 confirms that the motion of the particle is indeed the one described above. Moreover, the proof of Theorem 2.16 offers a way of approximating its trajectory.

Example 2.7 Let us have another look at the radioactive disintegration model we discussed in Sect. 1.3.1. If $x(t)$ denotes the quantity of radioactive material, then the evolution of this quantity is governed by the ODE

$$x'(t) = -\alpha x(t) + g(t), \tag{2.103}$$

where $g(t)$ denotes the amount of radioactive material that is added or extracted per unit of time at time t. Clearly, $x(t)$ is a solution of (2.103) only for those t such that $x(t) > 0$. That is why it is more appropriate to assume that x satisfies the following equations

$$\begin{aligned} x(t) &\geq 0, \;\; \forall t \geq 0, \\ x'(t) &= -\alpha x(t) + g(t), \;\; \forall t \in E_x \\ x'(t) &= \max\{g(t), 0\}, \;\; \forall t \in [0, \infty) \setminus E_x, \end{aligned} \tag{2.104}$$

where the set

$$E_x := \big\{ t \geq 0;\;\; x(t) > 0 \big\}$$

is also one of the unknowns in the above problem. This is a so-called "free boundary" problem. Let us observe that (2.104) is equivalent to the variational inequality (2.93) with

$$K := \big\{ x \in \mathbb{R};\;\; x \geq 0 \big\}.$$

More precisely, (2.104) is equivalent to

$$\big(x'(t) + \alpha x(t) - g(t) \big) \cdot \big(x(t) - y \big) \leq 0, \;\; \forall y \in K, \tag{2.105}$$

for almost all $t \geq 0$.

In formulation (2.105), the set E_x has disappeared, but we have to pay a price, namely, the new equation is a differential inclusion.

Example 2.8 Consider a factory consisting of n production units, each generating only one type of output. We denote by $x_i(t)$ the size of the output of unit i at time t, by $c_i(t)$ the demand for the product i at time t, and by $p_i(t)$ the rate at which the output i is produced. The demands and stocks define, respectively, the vector-valued maps

$$\boldsymbol{c}(t) := \begin{bmatrix} c_1(t) \\ \vdots \\ c_n(t) \end{bmatrix}, \quad \boldsymbol{x}(t) := \begin{bmatrix} x_1(t) \\ \vdots \\ x_n(t) \end{bmatrix},$$

and we will assume that the demand vector depends linearly on the stock vector, that is,

$$\boldsymbol{c}(t) = C\boldsymbol{x}(t) + \boldsymbol{d}(t),$$

where C is an $n \times n$ matrix, and $\boldsymbol{d} : [0, T] \to \mathbb{R}^n$ is a differentiable map. For $i = 1, \ldots, n$, define

$$E_{x_i} := \big\{ t \in [0, T];\;\; x_i(t) > 0 \big\}.$$

Obviously, the functions x_i satisfy the following equations

$$\begin{aligned} x_i(t) \geq 0, \;\; \forall t \in [0, T], \\ x_i'(t) = p_i(t) - \big(C\boldsymbol{x}(t)\big)_i - d_i(t), \;\; t \in E_{x_i}, \\ x_i'(t) - p_i(t) + \big(C\boldsymbol{x}(t)\big)_i + d_i(t) \geq 0, \;\; \forall t \in [0, T] \setminus E_{x_i}. \end{aligned} \tag{2.106}$$

We can now see that the solutions of the variational problem (2.92) and (2.93) with

$$\boldsymbol{f}(\boldsymbol{x}) = -C\boldsymbol{x}, \;\; \boldsymbol{g}(t) = \boldsymbol{p}(t) - \boldsymbol{d}(t),$$

and

$$K = \big\{ \boldsymbol{x} \in \mathbb{R}^n;\;\; x_i \geq 0, \;\; \forall i = 1, \ldots, n \big\},$$

are also solutions of (2.106).

Remark 2.8 Theorem 2.16 extends to differential inclusions of the form

$$\begin{aligned} &\boldsymbol{x}'(t) \in \boldsymbol{f}(\boldsymbol{x}(t)) + \phi(\boldsymbol{x}(t)) + \boldsymbol{g}(t), \;\; \text{a.e. } t \in (0, T), \\ &\boldsymbol{x}(0) = \boldsymbol{x}_0, \end{aligned} \tag{2.107}$$

where $\boldsymbol{f} : \mathbb{R}^n \to \mathbb{R}^n$ is Lipschitz and $\phi : D \subset \mathbb{R}^n \to 2^{\mathbb{R}^n}$ is a maximal dissipative mapping, that is,

$$(v_1 - v_2, u_1 - u_2) \leq 0, \;\; \forall v_i \in \phi(u_i), \;\; i = 1, 2,$$

and the range of the map $u \to u + \lambda\phi(u)$ is all of $\mathbb{R}^n$ for $\lambda > 0$. The method of proof is essentially the same as that of Theorem 2.16. Namely, one approximates (2.107) by

$$\begin{aligned} &\boldsymbol{x}'(t) = \boldsymbol{f}(\boldsymbol{x}(t)) + \phi_\varepsilon(\boldsymbol{x}(t)) + \boldsymbol{g}(t), \;\; t \in (0, T), \\ &\boldsymbol{x}(0) = \boldsymbol{x}_0, \end{aligned} \tag{2.108}$$

where ϕ_ε is the Lipschitz mapping $\frac{1}{\varepsilon}((I - \varepsilon\phi)^{-1}\boldsymbol{x} - \boldsymbol{x})$, $\varepsilon > 0$, $\boldsymbol{x} \in \mathbb{R}^n$.

Then, one obtains the solution to (2.107) as $\boldsymbol{x}(t) = \lim\limits_{\varepsilon \to 0} \boldsymbol{x}_\varepsilon(t)$, where $\boldsymbol{x}_\varepsilon \in C^1([0, T]; \mathbb{R}^n)$ is the solution to (2.108). We refer to [2] for details and more general results. We note, however, that this result applies to the Cauchy problem with discontinuous monotone functions ϕ. For instance, if ϕ is discontinuous in $\boldsymbol{x}^0$, then one fills the jump at $\boldsymbol{x}^0$ by redefining ϕ as

$$\widetilde{\phi}(\boldsymbol{x}) = \begin{cases} \phi(\boldsymbol{x}) & \text{for } \boldsymbol{x} \neq \boldsymbol{x}^0, \\ \lim\limits_{\boldsymbol{y} \to \boldsymbol{x}^0} \phi(\boldsymbol{y}) & \text{for } \boldsymbol{x} = \boldsymbol{x}^0. \end{cases}$$

Clearly, $\widetilde{\phi}$ is maximal dissipative.

Problems

2.1 Find the maximal interval of existence for the solution of the Cauchy problem

$$\begin{aligned} x' &= -x^2 + t + 1, \\ x(0) &= 1, \end{aligned}$$

and then find the first three Picard iterations of this problem.

2.2 Consider the Cauchy problem

$$\begin{aligned} x' &= f(t, x), \\ x(t_0) &= x_0, \quad (t_0, x_0) \in \Omega \subset \mathbb{R}^2, \end{aligned} \tag{2.109}$$

where the function f is continuous in (t, x) and locally Lipschitz in x. Prove that if $x_0 \geq 0$ and $f(t, 0) > 0$ for any $t \geq t_0$, then the saturated solution $x(t; t_0, x_0)$ is nonnegative for any $t \geq t_0$ in the existence interval.

2.3 Consider the system

$$\begin{aligned} \boldsymbol{x}' &= \boldsymbol{f}(t, \boldsymbol{x}), \\ \boldsymbol{x}(t_0) &= \boldsymbol{x}_0, \quad t_0 \geq 0, \end{aligned}$$

where the function $\boldsymbol{f} : [0, \infty) \times \mathbb{R}^n \to \mathbb{R}^n$ is continuous in $(t, \boldsymbol{x})$, locally Lipschitz in $\boldsymbol{x}$ and satisfies the condition

$$\big(\boldsymbol{f}(t, \boldsymbol{x}), P\boldsymbol{x} \big) \leq 0, \quad \forall t \geq 0, \quad \boldsymbol{x} \in \mathbb{R}^n, \tag{2.110}$$

where P is a real, symmetric and positive definite $n \times n$ matrix. Prove that any right-saturated solution of the system is defined on the semi-axis $[t_0, \infty)$.

Hint. Imitate the argument used in the proof of Theorem 2.12. Another approach is to replace the scalar product of $\mathbb{R}^n$ by

$$\langle u, v \rangle = (u, Pv), \ \forall u, v \in \mathbb{R}^n,$$

and argue as in the proof of Theorem 2.13.

2.4 Consider the Cauchy problem

$$x'' + ax + f(x') = 0, \quad x(t_0) = x_0, \quad x'(t_0) = x_1, \tag{2.111}$$

where a is a positive constant, and $f : \mathbb{R} \to \mathbb{R}$ is a locally Lipschitz function satisfying

$$yf(y) \geq 0, \quad \forall y \in \mathbb{R}.$$

Prove that any right-saturated solution of (2.111) is defined on the semi-axis $[t_0, \infty)$.

Hint. Multiplying (2.111) by x' we deduce

$$\frac{1}{2}\frac{d}{dt}|x'(t)|^2 + a|x'(t)|^2 \leq 0, \quad \forall t \geq t_0.$$

Conclude that the functions $x(t)$ and $x'(t)$ are bounded and then use Theorem 2.11.

2.5 In the anisotropic theory of relativity due to V.G. Boltyanski, the propagation of light in a neighborhood of a mass m located at the origin of $\mathbb{R}^3$ is described by the equation

$$\boldsymbol{x}' = -\frac{m\gamma}{\|\boldsymbol{x}\|_{\mathbf{e}}^3}\boldsymbol{x} + \boldsymbol{u}(t), \tag{2.112}$$

where γ is a positive constant, $\boldsymbol{u} : [0, \infty) \to \mathbb{R}^3$ is a continuous and *bounded* function, that is,

$$\exists C > 0; \quad \|\boldsymbol{u}(t)\|_{\mathbf{e}} \leq C, \quad \forall t \geq 0,$$

and $\boldsymbol{x}(t) \in \mathbb{R}^3$ is the location of the photon at time t. Prove that there exists an $r > 0$ such that all the trajectories of (2.112), which start at $t = 0$ in the ball

$$B_r := \{\boldsymbol{x} \in \mathbb{R}^3; \quad \|\boldsymbol{x}\|_{\mathbf{e}} < r\},$$

will stay inside the ball as long as they are defined. (Such a ball is called a *black hole* in astrophysics.)

Hint. Take the scalar product of (2.112) with $\boldsymbol{x}(t)$ to obtain the differential inequality

$$\frac{1}{2}\frac{d}{dt}\|\boldsymbol{x}(t)\|_{\mathbf{e}}^2 = -\frac{m\gamma}{\|\boldsymbol{x}(t)\|_{\mathbf{e}}} + \big(\boldsymbol{u}(t), \boldsymbol{x}(t)\big) \leq -\frac{m\gamma}{\|\boldsymbol{x}(t)\|_{\mathbf{e}}} + C\|\boldsymbol{x}(t)\|_{\mathbf{e}}.$$

Use this differential inequality to obtain an upper estimate for $\|\boldsymbol{x}(t)\|_{\mathbf{e}}$.

2.6 (Wintner's extendibility test) Prove that, if the continuous function $f = f(t, x) : \mathbb{R} \times \mathbb{R} \to \mathbb{R}$ is locally Lipschitz in x and satisfies the inequality

$$|f(t, x)| \leq \mu(|x|), \quad \forall (t, x) \in \mathbb{R} \times \mathbb{R}, \tag{2.113}$$

where $\mu : (0, \infty) \to (0, \infty)$ satisfies

$$\int_0^\infty \frac{dr}{\mu(r)} < \infty, \tag{2.114}$$

then all the solutions of $x' = f(t, x)$ are defined on the whole axis $\mathbb{R}$.

Hint. According to Theorem 2.11, it suffices to show that all the solutions of $x' = f(t, x)$ are bounded. To prove this, we conclude from (2.113) that

$$\left| \int_{x_0}^{x(t)} \frac{dr}{\mu(r)} \right| \leq |t - t_0|, \quad \forall t,$$

and then invoke (2.114).

2.7 Prove that the saturated solution of the Cauchy problem

$$\begin{aligned} x' &= e^{-x^2} + t^2, \\ x(0) &= 1, \end{aligned} \tag{2.115}$$

is defined on the interval $\left[0, \frac{1}{2}\right]$. Use Euler's method with step size $h = 10^{-2}$ to find an approximation of this solution at the nodes $t_j = jh$, $j = 1, \ldots, 50$.

2.8 Let $f : \mathbb{R} \to \mathbb{R}$ be a continuous and nonincreasing function. Consider the Cauchy problem

$$\begin{aligned} x'(t) &= f(x), \quad \forall t \geq 0, \\ x(0) &= x_0. \end{aligned} \tag{2.116}$$

According to Theorem 2.13, this problem has a unique solution $x(t)$ which exists on $[0, \infty)$.

(a) Prove that, for any $\lambda > 0$, the function

$$\mathbb{1} - \lambda f : \mathbb{R} \to \mathbb{R}, \; x \mapsto x - \lambda f(x),$$

is bijective. For any integer $n > 0$, we set

$$(\mathbb{1} - \lambda f)^{-n} := \underbrace{(\mathbb{1} - \lambda f)^{-1} \circ \cdots \circ (\mathbb{1} - \lambda f)^{-1}}_{n}.$$

(b) Prove that $x(t)$ is given by the formula

$$x(t) = \lim_{n \to \infty} \left(\mathbb{1} - \frac{t}{n} f \right)^{-n} x_0, \quad \forall t \geq 0. \tag{2.117}$$

Hint. Fix $t > 0$, $n > 0$, set

$$h_n := \frac{t}{n}$$

and define iteratively

$$x_0^n = x_0, \quad \frac{x_i^n - x_{i-1}^n}{h_n} = f(x_i^n), \quad i = 1, \dots n,$$

that is,

$$x_i^n = \left(\mathbb{1} - \frac{t}{n}f\right)^{-1} x_{i-1}^n = \left(\mathbb{1} - \frac{t}{n}f\right)^{-i} x_0. \tag{2.118}$$

Let $x^n : [0, t] \to \mathbb{R}$ be the unique continuous function which is linear on each of the intervals $[(i-1)h_n, ih_n]$ and satisfies

$$x^n(ih_n) = x_i^n, \quad \forall i = 0, \dots, n.$$

Argue as in the proof of Peano's theorem to show that x^n converges uniformly to x on $[0, t]$ as $n \to \infty$. Equality (2.117) now follows from (2.118).

2.9 Consider the Cauchy problem

$$\begin{aligned} x' &= f(x), \quad t \geq 0 \\ x(0) &= x_0, \end{aligned} \tag{2.119}$$

where $f : \mathbb{R} \to \mathbb{R}$ is a continuous nonincreasing function. Let $x = \varphi(t)$ be a solution of (2.119). Prove that, if the set

$$F := \{ y \in \mathbb{R}; \quad f(y) = 0 \}$$

is nonempty, then the following hold.

(i) The function $t \mapsto |x'(t)|$ is nonincreasing on $[0, \infty)$.
(ii) $\lim_{t\to\infty} |x'(t)| = 0$.
(iii) There exists an $x_\infty \in F$ such that $\lim_{t\to\infty} x(t) = x_\infty$.

Hint. Since f is nonincreasing we have

$$\begin{aligned} \frac{1}{2}\frac{d}{dt}\big(x(t+h) - x(t)\big)^2 &= \big(x'(t+h) - x'(t)\big)\big(x(t+h) - x(t)\big) \\ &= \big(f(x(t+h)) - f(x(t)\big)\big(x(t+h) - x(t)\big) \leq 0. \end{aligned}$$

Hence, for any $h \geq 0$ and $t_2 \geq t_1 \geq 0$, we have

$$\big| x(t_2 + h) - x(t_1) \big| \leq \big| x(t_1 + h) - x(t_1) \big|.$$

This proves (i). To prove (ii) multiply both sides of (2.119) by $x(t) - y_0$, where $y_0 \in F$. Conclude, similarly, that

$$\frac{d}{dt}\big(x(t) - y_0\big)^2 \le 0,$$

showing that $\lim_{t\to\infty}(x(t) - y_0)^2$ exists. Next multiply the equation by $x'(t)$ to obtain

$$\int_0^t |x'(s)|^2 ds = g(x(t)) - g(x_0),$$

where g is an antiderivative of f. We deduce that

$$\int_0^\infty |x'(t)|^2 dt < \infty,$$

which when combined with (i) yields (ii). Next, pick a subsequence $t_n \to \infty$ such that $x(t_n) \to y_0$. From (i) and (ii), it follows that $y_0 \in F$. You can now conclude that

$$\lim_{t\to\infty} x(t) = y_0.$$

2.10 Prove that the conclusions of Problem 2.9 remain true for the system

$$\begin{aligned} \boldsymbol{x}' &= \boldsymbol{f}(\boldsymbol{x}), \\ \boldsymbol{x}(0) &= \boldsymbol{x}_0, \end{aligned}$$

where $\boldsymbol{f} : \mathbb{R}^n \to \mathbb{R}^n$ is a dissipative and continuous mapping of the form $\boldsymbol{f} = \nabla \boldsymbol{g}$, where $\boldsymbol{g} : \mathbb{R}^n \to \mathbb{R}$ is of class C^1 and $\boldsymbol{g} \ge 0$.

Hint. One proceeds as above by taking into account that

$$\frac{d}{dt} g(\boldsymbol{x}(t)) = (\boldsymbol{x}'(t), \boldsymbol{f}(\boldsymbol{x}(t))),\ \forall t \ge 0.$$

2.11 Consider the system

$$\begin{aligned} \boldsymbol{x}' &= \boldsymbol{f}(\boldsymbol{x}) - \lambda \boldsymbol{x} + \boldsymbol{f}_0, \\ \boldsymbol{x}(0) &= \boldsymbol{x}_0, \end{aligned}$$

where $\boldsymbol{f} : \mathbb{R}^n \to \mathbb{R}^n$ is continuous and dissipative, $\lambda > 0$ and $\boldsymbol{f}_0, \boldsymbol{x}_0 \in \mathbb{R}^n$. Prove that

(a) $\lim_{t\to\infty} \boldsymbol{x}(t)$ exists (call this limit $\boldsymbol{x}_\infty$),
(b) $\lim_{t\to\infty} \boldsymbol{x}'(t) = 0$,
(c) $\lambda \boldsymbol{x}_\infty - \boldsymbol{f}(\boldsymbol{x}_\infty) = f_0$.

Hint. For each $h > 0$, one has

$$\frac{1}{2}\frac{d}{dt}\|\boldsymbol{x}(t+h) - \boldsymbol{x}(t)\|_{\mathbf{e}}^2 + \lambda\|\boldsymbol{x}(t+h) - \boldsymbol{x}(t)\|_{\mathbf{e}}^2 \le 0,$$

which implies that (a) holds and that

$$\|\boldsymbol{x}'(t)\|_{\mathbf{e}} \leq e^{-\lambda t}\|\boldsymbol{x}'(0)\|_{\mathbf{e}}, \quad \forall t \geq 0.$$

2.12 Prove that (2.117) remains true for solutions $\boldsymbol{x}$ to the Cauchy problem (2.116), where $\boldsymbol{f} : \mathbb{R}^n \to \mathbb{R}^n$ is continuous and dissipative.

Hint. By Problem 2.11(c), the function $\mathbb{1} - \lambda \boldsymbol{f} : \mathbb{R}^n \to \mathbb{R}^n$ is bijective and the Euler scheme (2.38) is equivalent to

$$\boldsymbol{x}_i^n = \left(\mathbb{1} - \frac{t}{k}\boldsymbol{f}\right)^{-i} \boldsymbol{x}_0 \text{ for } t \in (ih_k, (i+1)h_k).$$

Then, by the convergence of this scheme, one obtains

$$\boldsymbol{x}(t) = \lim_{k\to\infty}\left(\mathbb{1} - \frac{t}{k}\right)^{-k} \boldsymbol{x}_0, \quad \forall t \geq 0.$$

2.13 Consider the equation $x' = f(t, x)$, where $f : \mathbb{R}^2 \to \mathbb{R}$ is a function continuous in (t, x) and locally Lipschitz in x, and satisfies the growth constraint

$$\big| f(t, x) \big| \leq \alpha(t)|x|, \quad \forall (t, x) \in \mathbb{R}^2,$$

where

$$\int_{t_0}^{\infty} \alpha(t)dt < \infty.$$

(a) Prove that any solution of the equation has finite limit as $t \to \infty$.

(b) Prove that if, additionally, f satisfies the Lipschitz condition

$$\big| f(t, x) - f(t, y) \big| \leq \alpha(t)|x - y|, \quad \forall t \in \mathbb{R}, \quad x, y \in \mathbb{R},$$

then there exists a bijective correspondence between the initial values of the solutions and their limits at ∞.

Hint. Use Theorem 2.11 as in Example 2.1.

2.14 Prove that the maximal existence interval of the Cauchy problem

$$\begin{aligned} x' &= ax^2 + t^2, \\ x(0) &= x_0, \end{aligned} \tag{2.120}$$

(a is a positive constant) is bounded from above. Compare this with the situation encountered in Example 2.2.

Hint. Let $x_0 \geq 0$ and $[0, T)$ be the maximal interval of definition on the right. Then, on this interval,

$$x(t; 0, x_0) \geq x_0 + \frac{t^3}{3}, \quad \frac{1}{x(t; 0, x_0)} \leq \frac{1}{x_0} - at.$$

Hence, $T = (ax_0)^{-1}$.

2.15 Consider the Volterra integral equation

$$\boldsymbol{x}(t) = \boldsymbol{g}(t) + \int_a^t \boldsymbol{f}(s, \boldsymbol{x}(s))ds, \quad t \in [a, b],$$

where $\boldsymbol{g} \in C([a, b]; \mathbb{R}^n)$, $\boldsymbol{f} : [a, b] \times \mathbb{R}^n \to \mathbb{R}^n$ is continuous and

$$\|\boldsymbol{f}(s, \boldsymbol{x}) - \boldsymbol{f}(s, \boldsymbol{y})\| \leq L\|\boldsymbol{x} - \boldsymbol{y}\|, \forall s \in [a, b], \boldsymbol{x} \in \mathbb{R}^n.$$

Prove that there is a unique solution $\boldsymbol{x} \in C([a, b]; \mathbb{R}^n)$.

Hint. One proceeds as in the proofs of Theorems 2.1 and 2.4 by the method of successive approximations

$$\boldsymbol{x}_{n+1}(t) = \boldsymbol{g}(t) + \int_0^t f(s, \boldsymbol{x}_n(s))ds, \quad t \in [a, b],$$

and proving that $\{\boldsymbol{x}_n\}$ is uniformly convergent.

Chapter 3
Systems of Linear Differential Equations

The study of linear systems of ODEs offers an example of a well-put-together theory, based on methods and results from linear algebra. As we will see, there exist many similarities between the theory of systems of linear algebraic equations and the theory of systems of linear ODEs. In applications, linear systems appear most often as "first approximations" of more complex processes.

3.1 Notation and Some General Results

A system of first-order linear ODEs has the form

$$x_i'(t) = \sum_{j=1}^{n} a_{ij}(t)x_j(t) + b_i(t), \quad i = 1, \dots, n, \quad t \in I, \tag{3.1}$$

where I is an interval of the real axis and $a_{ij}, b_i : I \to \mathbb{R}$ are continuous functions. System (3.1) is called *nonhomogeneous*. If $b_i(t) \equiv 0$, $\forall i$, then the system is called *homogeneous*. In this case, it has the form

$$x_i'(t) = \sum_{j=1}^{n} a_{ij}(t)x_i(t), \quad i = 1, \dots, n, \quad t \in I. \tag{3.2}$$

Using the vector notation as in Sect. 2.2, we can rewrite (3.1) and (3.2) in the form

$$\boldsymbol{x}'(t) = A(t)\boldsymbol{x}(t) + \boldsymbol{b}(t), \quad t \in I \tag{3.3}$$

$$\boldsymbol{x}'(t) = A(t)\boldsymbol{x}(t), \quad t \in I, \tag{3.4}$$

V. Barbu, *Differential Equations*, Springer Undergraduate Mathematics Series,
DOI 10.1007/978-3-319-45261-6_3

where

$$x(t) := \begin{bmatrix} x_1(t) \\ \vdots \\ x_n(t) \end{bmatrix}, \quad b(t) := \begin{bmatrix} b_1(t) \\ \vdots \\ b_n(t) \end{bmatrix},$$

and A is the $n \times n$ matrix $A := \big(a_{ij}(t) \big)_{1 \le i,j \le n}$.

Obviously, the local existence and uniqueness theorem (Theorem 2.2), as well as the results concerning global existence and uniqueness apply to system (3.3). Thus, for any $t_0 \in I$, and any $\boldsymbol{x}_0 \in \mathbb{R}^n$, there exists a unique saturated solution of (3.1) satisfying the initial condition

$$\boldsymbol{x}(t_0) = \boldsymbol{x}_0. \tag{3.5}$$

In this case, the domain of existence of the saturated solution coincides with the interval I. In other words, we have the following result.

Theorem 3.1 *The saturated solution $\boldsymbol{x} = \varphi(t)$ of the Cauchy problem* (3.1) and (3.5) *is defined on the entire interval I.*

Proof Let $(\alpha, \beta) \subset I = (t_1, t_2)$ be the interval of definition of the solution $\boldsymbol{x} = \varphi(t)$. According to Theorem 2.10 applied to system (3.3), $\Omega = (t_1, t_2) \times \mathbb{R}^n, \boldsymbol{f}(t, \boldsymbol{x}) = A(t)\boldsymbol{x}$, if $\beta < t_2$, or $\alpha > t_1$, then the function $\varphi(t)$ is unbounded in the neighborhood of β, respectively α. Suppose that $\beta < t_2$. From the integral identity

$$\varphi(t) = \boldsymbol{x}_0 + \int_{t_0}^{t} A(s)\varphi(s)ds + \int_{t_0}^{t} \boldsymbol{b}(s)ds, \quad t_0 \le t < \beta,$$

we obtain, by passing to norms,

$$\|\varphi(t)\| \le \boldsymbol{x}_0 + \int_{t_0}^{t} \|A(s)\| \cdot \|\varphi(s)\|ds + \int_{t_0}^{t} \|\boldsymbol{b}(s)\|ds, \quad t_0 \le t < \beta. \tag{3.6}$$

The functions $t \mapsto \|A(t)\|$ and $t \mapsto \|\boldsymbol{b}(t)\|$ are continuous on $[t_0, t_2)$ and thus are bounded on $[t_0, \beta]$. Hence, there exists an $M > 0$ such that

$$\|A(t)\| + \|\boldsymbol{b}(t)\| \le M, \quad \forall t \in [t_0, \beta]. \tag{3.7}$$

From inequalities (3.6) and (3.7) and Gronwall's lemma, we deduce that

$$\|\varphi(t)\| \le \big(\boldsymbol{x}_0 + (\beta - t_0)M\big)e^{(\beta - t_0)M}, \quad \forall t \in [t_0, \beta).$$

We have reached a contradiction which shows that, necessarily, $\beta = t_2$. The equality $\alpha = t_1$ can be proven in a similar fashion. This completes the proof of Theorem 3.1. □

3.2 Homogeneous Systems of Linear Differential Equations

In this section, we will investigate system (3.2) (equivalently, (3.4)). We begin with a theorem on the structure of the set of solutions.

Theorem 3.2 *The set of solutions of system* (3.2) *is a real vector space of dimension n.*

Proof The set of solutions is obviously a real vector space. Indeed, the sum of two solutions of (3.2) and the multiplication by a scalar of a solution are also solutions of this system.

To prove that the dimension of this vector space is n, we will show that there exists a linear isomorphism between the space E of solutions of (3.2) and the space $\mathbb{R}^n$. Fix a point $t_0 \in I$ and denote by Γ_{t_0} the map $E \to \mathbb{R}^n$ that associates to a solution $\boldsymbol{x} \in E$ its value at $t_0 \in E$, that is,

$$\Gamma_{t_0}(\boldsymbol{x}) = \boldsymbol{x}(t_0) \in \mathbb{R}^n, \quad \forall \boldsymbol{x} \in E.$$

The map Γ_{t_0} is obviously linear. The existence and uniqueness theorem concerning the Cauchy problems associated with (3.2) implies that Γ_{t_0} is also surjective and injective. This completes the proof of Theorem 3.2. □

The above theorem shows that the space E of solutions of (3.2) admits a basis consisting of n solutions. Let

$$\{\boldsymbol{x}^1, \boldsymbol{x}^2, \dots, \boldsymbol{x}^n\}$$

be one such basis. In particular, $\boldsymbol{x}^1, \boldsymbol{x}^2, \dots, \boldsymbol{x}^n$ are n linearly independent solutions of (3.2), that is, the only constants $c_1, c_2, \dots, c_n$ such that

$$c_1\boldsymbol{x}^1(t) + c_2\boldsymbol{x}^2(t) + \cdots + c_n\boldsymbol{x}^n(t) = 0, \quad \forall t \in I,$$

are the null ones, $c_1 = c_2 = \cdots = c_n = 0$. The matrix $\boldsymbol{X}(t)$ whose columns are given by the function $\boldsymbol{x}^1(t), \boldsymbol{x}^2(t), \dots, \boldsymbol{x}^n(t)$,

$$\boldsymbol{X}(t) := \big[\boldsymbol{x}^1(t), \boldsymbol{x}^2(t), \dots, \boldsymbol{x}^n(t)\big], \quad t \in I,$$

is called a *fundamental matrix*. It is easy to see that the matrix $\boldsymbol{X}(t)$ is a solution to the differential equation

$$\boldsymbol{X}'(t) = A(t)\boldsymbol{X}(t), \quad t \in I, \tag{3.8}$$

where we denote by $\boldsymbol{X}'(t)$ the matrix whose entries are the derivatives of the corresponding entries of $\boldsymbol{X}(t)$.

A fundamental matrix is not unique. Let us observe that any matrix $\boldsymbol{Y}(t) = \boldsymbol{X}(t)C$, where C is a constant nonsingular matrix, is also a fundamental matrix of (3.2). Conversely, any fundamental matrix $\boldsymbol{Y}(t)$ of system (3.2) can be represented as a product

$$Y(t) = X(t)C, \quad \forall t \in I,$$

where C is a constant, nonsingular $n \times n$ matrix. This follows from the next simple result.

Corollary 3.1 *Let $X(t)$ be a fundamental solution of the system* (3.2). *Then any solution $\boldsymbol{x}(t)$ of* (3.2) *has the form*

$$\boldsymbol{x}(t) = \boldsymbol{X}(t)\boldsymbol{c}, \quad t \in I, \tag{3.9}$$

where $\boldsymbol{c}$ is a vector in $\mathbb{R}^n$ (that depends on $\boldsymbol{x}(t)$).

Proof Equality (3.9) follows from the fact that the columns of $\boldsymbol{X}(t)$ form a basis of the space of solutions of system (3.2). Equality (3.9) simply states that any solution of (3.2) is a linear combinations of solutions forming a basis for the space of solutions. □

Given a collection $\{\boldsymbol{x}^1, \ldots, \boldsymbol{x}^n\}$ of solutions of (3.2), we define the *Wronskian* of this collection to be the determinant

$$W(t) := \det \boldsymbol{X}(t), \tag{3.10}$$

where $\boldsymbol{X}(t)$ denotes the matrix with columns $\boldsymbol{x}(y), \ldots, \boldsymbol{x}^n(t)$. The next result, due to the Polish mathematician *H. Wronski* (1778–1853), explains the relevance of the quantity $W(t)$.

Theorem 3.3 *The collection of solutions $\{\boldsymbol{x}^1, \ldots, \boldsymbol{x}^n\}$ of* (3.2) *is linearly independent if and only if its Wronskian $W(t)$ is nonzero at a point of the interval I (equivalently, on the entire interval I).*

Proof Clearly, if the collection is linearly dependent, then

$$\det \boldsymbol{X}(t) = W(t) = 0, \quad \forall t \in I,$$

where $\boldsymbol{X}(t)$ is the matrix $[\boldsymbol{x}^1(t), \ldots, \boldsymbol{x}^n(t)]$.

Conversely, suppose that the collection is linearly independent. We argue by contradiction, and assume that $W(t_0) = 0$ at some $t_0 \in I$. Consider the linear homogeneous system

$$\boldsymbol{X}(t_0)\boldsymbol{x} = 0. \tag{3.11}$$

Since $\det \boldsymbol{X}(t_0) = W(t_0) = 0$, system (3.11) admits a nontrivial solution $\boldsymbol{c}_0 \in \mathbb{R}^n$. The function $\boldsymbol{y}(t) := \boldsymbol{X}(t)\boldsymbol{c}_0$ is obviously a solution of (3.2) vanishing at t_0. The uniqueness theorem implies

$$\boldsymbol{y}(t) = 0, \quad \forall t \in I.$$

Hence $\boldsymbol{X}(t)\boldsymbol{c}_0 = 0, \forall t \in I$, which contradicts the linear independence of the collection $\{\boldsymbol{x}^1, \ldots, \boldsymbol{x}^n\}$. This completes the proof of the theorem. □

Theorem 3.4 (J. Liouville (1809–1882)) *Let $W(t)$ be the Wronskian of a collection of n solutions of system* (3.2). *Then we have the equality*

$$W(t) = W(t_0)\exp\left(\int_{t_0}^{t} \operatorname{tr} A(s)ds\right), \quad \forall t_0, t \in I, \tag{3.12}$$

where $\operatorname{tr} A(t)$ *denotes the trace of the matrix* $A(t)$,

$$\operatorname{tr} A(t) = \sum_{i=1}^{n} a_{ii}(t).$$

Proof Without loss of generality, we can assume that $W(t)$ is the Wronskian of a linearly independent collection of solutions $\{\boldsymbol{x}^1, \dots, \boldsymbol{x}^n\}$. (Otherwise, equality (3.12) would follow trivially from Theorem 3.3.) Denote by $\boldsymbol{X}(t)$ the fundamental matrix with columns $\boldsymbol{x}^1(t), \dots, \boldsymbol{x}^n(t)$.

From the definition of the derivative, we deduce that for any $t \in I$ we have

$$\boldsymbol{X}(t+\varepsilon) = \boldsymbol{X}(t) + \varepsilon \boldsymbol{X}'(t) + o(\varepsilon), \quad \text{as } \varepsilon \to 0.$$

From (3.8), it follows that

$$\boldsymbol{X}(t+\varepsilon) = X(t) + \varepsilon A(t)\boldsymbol{X}(t) + o(\varepsilon), \quad \forall t \in I. \tag{3.13}$$

In (3.13), we take the determinant of both sides and we find that

$$W(t+\varepsilon) = \det\big(\mathbb{1} + A(t) + o(\varepsilon)\big)W(t) = \big(1 + \varepsilon \operatorname{tr} A(t) + o(\varepsilon)\big).$$

Letting $\varepsilon \to 0$ in the above equality, we obtain

$$W'(t) = \big(\operatorname{tr} A(t)\big)W(t).$$

Integrating the above linear differential equation, we get (3.12), as claimed. □

Remark 3.1 From Liouville's theorem, we deduce in particular the fact that, if the Wronskian is nonzero at a point, then it is nonzero everywhere.

Taking into account that the determinant of a matrix is the oriented volume of the parallelepiped determined by its columns, Liouville's formula (3.12) describes the variation in time of the volume of the parallelepiped determined by $\{\boldsymbol{x}^1(t), \dots, \boldsymbol{x}^n(t)\}$. In particular, if $\operatorname{tr} A(t) = 0$, then this volume is conserved along the trajectories of system (3.2).

This fact admits a generalization to nonlinear differential systems of the form

$$\boldsymbol{x}' = \boldsymbol{f}(t, \boldsymbol{x}), \quad t \in I, \tag{3.14}$$

where $f : I \times \mathbb{R}^n \to \mathbb{R}^n$ is a C^1-map such that

$$\operatorname{div}_x f(t, x) \equiv 0. \tag{3.15}$$

Assume that for any $x_0 \in \mathbb{R}^n$ there exists a solution $S(t; t_0 x_0) = x(t; t_0, x_0)$ of system (3.14) satisfying the initial condition $x(t_0) = x_0$ and defined on the interval I. Let D be a domain in $\mathbb{R}^n$ and set

$$D(t) = S(t)D := \big\{ S(t; t_0, x_0); \;\; x_0 \in D \big\}.$$

Liouville's theorem from statistical physics states that *the volume of $D(t)$ is constant.* The proof goes as follows.

For any $t \in I$, we have

$$\operatorname{Vol} D(t) = \int_D \det \frac{\partial S(t, x)}{\partial x} \, dx. \tag{3.16}$$

On the other hand, Theorem 3.14 to come shows that $\frac{\partial S(t,x)}{\partial x}$ is the solution of the Cauchy problem

$$\frac{d}{dt} \frac{\partial S(t, x)}{\partial x} = \frac{\partial f}{\partial x}\big(t, S(t, x)\big) \frac{\partial S(t, x)}{\partial x}$$

$$\frac{\partial S(t_0, x)}{\partial x} = \mathbb{1}.$$

We deduce that

$$\frac{\partial S(t, x)}{\partial x} = \mathbb{1} + (t - t_0) \frac{\partial f}{\partial x}(t_0, x) + o(t - t_0).$$

Hence

$$\begin{aligned} \det \frac{\partial S(t, x)}{\partial x} &= 1 + (t - t_0) \operatorname{tr} \frac{\partial f}{\partial x}(t_0, x) + o(t - t_0) \\ &= 1 + (t - t_0) \operatorname{div}_x f(t_0, x) + o(t - t_0). \end{aligned}$$

This yields

$$\operatorname{Vol} D(t) = \operatorname{Vol} D(t - t_0) + o(t - t_0).$$

In other words, $\big(\operatorname{Vol} D(t)\big)' \equiv 0$, and thus $\operatorname{Vol} D(t)$ is constant.

In particular, Liouville's theorem applies to the Hamiltonian systems

$$\frac{\partial x}{\partial t} = \frac{\partial H}{\partial p}(x, p), \quad \frac{dp}{dt} = -\frac{\partial H}{\partial x}(x, p),$$

where $H : \mathbb{R}^n \times \mathbb{R}^n \to \mathbb{R}$ is a C^1-function.

3.3 Nonomogeneous Systems of Linear Differential Equations

In this section, we will investigate the nonhomogeneous system (3.1) or, equivalently, system (3.3). Our first result concerns the structure of the set of solutions.

Theorem 3.5 *Let $X(t)$ be a fundamental solution of the homogeneous system* (3.2) *and $\widetilde{\boldsymbol{x}}(t)$ a given solution of the nonhomogeneous system* (3.3). *Then the general solution of system* (3.3) *has the form*

$$\boldsymbol{x}(t) = \boldsymbol{X}(t)\boldsymbol{c} + \widetilde{\boldsymbol{x}}(t), \quad t \in I, \tag{3.17}$$

where $\boldsymbol{c}$ is an arbitrary vector in $\mathbb{R}^n$.

Proof Obviously, any function $\boldsymbol{x}(t)$ of the form (3.17) is a solution of (3.3). Conversely, let $\boldsymbol{y}(t)$ be an arbitrary solution of system (3.3) determined by its initial condition $\boldsymbol{y}(t_0) = \boldsymbol{y}_0$, where $t_0 \in I$ and $\boldsymbol{y}_0 \in \mathbb{R}^n$. Consider the linear algebraic system

$$\boldsymbol{X}(t_0)\boldsymbol{c} = \boldsymbol{y}_0 - \widetilde{\boldsymbol{x}}(t_0).$$

Since $\det \boldsymbol{X}(t_0) \neq 0$, the above system has a unique solution $\boldsymbol{c}_0$. Then the function $\boldsymbol{X}(t_0)\boldsymbol{c}_0 + \widetilde{\boldsymbol{x}}(t)$ is a solution of (3.3) and has the value $\boldsymbol{y}_0$ at t_0. The existence and uniqueness theorem then implies that

$$\boldsymbol{y}(t) = \boldsymbol{X}(t_0)\boldsymbol{c}_0 + \widetilde{\boldsymbol{x}}(t). \ \forall t \in I.$$

In other words, the arbitrary solution $\boldsymbol{y}(t)$ has the form (3.17). □

The next result clarifies the statement of Theorem 3.5 by offering a representation formula for a particular solution of (3.3).

Theorem 3.6 (Variation of constants formula) *Let $\boldsymbol{X}(t)$ be a fundamental matrix for the homogeneous system* (3.2). *Then the general solution of the nonhomogeneous system* (3.3) *admits the integral representation*

$$\boldsymbol{x}(t) = \boldsymbol{X}(t)\boldsymbol{c} + \int_{t_0}^{t} \boldsymbol{X}(t)\boldsymbol{X}(s)^{-1}\boldsymbol{b}(s)ds, \quad t \in I, \tag{3.18}$$

where $t_0 \in I$, $\boldsymbol{c} \in \mathbb{R}^n$.

Proof We seek a particular solution $\widetilde{\boldsymbol{x}}(t)$ of (3.3) of the form

$$\widetilde{\boldsymbol{x}}(t) = \boldsymbol{X}(t)\gamma(t), \quad t \in I, \tag{3.19}$$

where $\gamma : I \to \mathbb{R}^n$ is a function to be determined. Since $\widetilde{\boldsymbol{x}}(t)$ is supposed to be a solution of (3.3), we have

$$\boldsymbol{X}'(t)\gamma(t) + \boldsymbol{X}(t)\gamma'(t) = A(t)\boldsymbol{X}(t) + \boldsymbol{b}(t).$$

Using equality (3.8), we have $\boldsymbol{X}'(t) = A(t)\boldsymbol{X}(t)$ and we deduce that

$$\gamma'(t) = \boldsymbol{X}(t)^{-1}\boldsymbol{b}(t), \quad \forall t \in I,$$

and thus we can choose $\gamma(t)$ of the form

$$\gamma(t) = \int_{t_0}^{t} \boldsymbol{X}(s)^{-1}\boldsymbol{b}(s)ds, \quad t \in I, \tag{3.20}$$

where t_0 is some fixed point in I. The representation formula (3.18) now follows from (3.17), (3.19) and (3.20). □

Remark 3.2 From (3.18), it follows that the solution of system (3.3) satisfying the Cauchy condition $\boldsymbol{x}(t_0) = \boldsymbol{x}_0$ is given by the formula

$$\boldsymbol{x}(t) = \boldsymbol{X}(t)\boldsymbol{X}(t_0)^{-1}\boldsymbol{x}_0 + \int_{t_0}^{t} \boldsymbol{X}(t)\boldsymbol{X}(s)^{-1}\boldsymbol{b}(s)ds, \quad t \in I. \tag{3.21}$$

In mathematical systems theory the matrix $U(t,s) := \boldsymbol{X}(t)\boldsymbol{X}(s)^{-1}$, $s, t \in I$, is often called the *transition matrix*.

3.4 Higher Order Linear Differential Equations

Consider the linear homogeneous differential equation of order n

$$x^{(n)} + a_1(t)x^{(n-1)}(t) + \cdots + a_n(t)x(t) = 0, \quad t \in I, \tag{3.22}$$

and the associated nonhomogeneous equation

$$x^{(n)} + a_1(t)x^{(n-1)}(t) + \cdots + a_n(t)x(t) = f(t), \quad t \in I, \tag{3.23}$$

where a_i, $i = 1, \dots, n$, and f are continuous functions on an interval I.

Using the general procedure of reducing a higher order ODE to a system of first-order ODEs, we set

$$x_1 := x, \quad x_2 := x', \dots, x_n := x^{(n-1)}.$$

The homogeneous equation (3.22) is equivalent to the first-order linear differential system

$$
\begin{aligned}
x_1' &= x_2 \\
x_2' &= x_3 \\
\vdots &\quad \vdots \\
x_n' &= -a_n x_1 - a_{n-1} x_2 - \cdots - a_1 x_n .
\end{aligned}
\tag{3.24}
$$

In other words, the map Λ defined by

$$
x \mapsto \Lambda x := \begin{bmatrix} x \\ x' \\ \vdots \\ x^{(n-1)} \end{bmatrix}
$$

defines a linear isomorphism between the set of solutions of (3.22) and the set of solutions of the linear system (3.24). From Theorem 3.2, we deduce the following result.

Theorem 3.7 *The set of solutions to* (3.22) *is a real vector space of dimension* n.

Let us fix a basis $\{x_1, \ldots, x_n\}$ of the space of solutions to (3.22).

Corollary 3.2 *The general solution to* (3.22) *has the form*

$$
x(t) = c_1 x_1(t) + \cdots + c_n x_n(t), \tag{3.25}
$$

where $c_1, \ldots, c_n$ *are arbitrary constants.*

Just as in the case of linear differential systems, a collection of n linearly independent solutions of (3.22) is called a *fundamental* system (or collection) of solutions.

Using the isomorphism Λ, we can define the concept of the Wronskian of a collection of n solutions of (3.22). If $\{x_1, \ldots, x_n\}$ is such a collection, then its *Wronskian* is the function $W : I \to \mathbb{R}$ defined by

$$
W(t) := \det \begin{bmatrix} x_1 & \cdots\cdots & x_n \\ x_1' & \cdots\cdots & x_n' \\ \vdots & \vdots \; \vdots & \vdots \\ x_1^{(n-1)} & \cdots\cdots & x_n^{(n-1)} \end{bmatrix} . \tag{3.26}
$$

Theorem 3.3 has the following immediate consequence.

Theorem 3.8 *The collection of solutions* $\{x_1, \ldots, x_n\}$ *to* (3.22) *is fundamental if and only if its Wronskian is nonzero at a point or, equivalently, everywhere on* I.

Taking into account the special form of the matrix $A(t)$ corresponding to the system (3.24), we have the following consequence of Liouville's theorem.

Theorem 3.9 *For any $t_0, t \in I$ we have*

$$W(t) = W(t_0) \exp\left(-\int_{t_0}^{t} a_1(s)ds \right), \tag{3.27}$$

where $W(t)$ is the Wronskian of a collection of solutions.

Theorem 3.5 shows that the general solution of the nonhomogeneous equation (3.23) has the form

$$x(t) = c_1 x_1(t) + \cdots + c_n x_n(t) + \widetilde{x}(t), \tag{3.28}$$

where $\{x_1, \ldots, x_n\}$ is a fundamental collection of solutions of the homogeneous equation (3.22), and $\widetilde{x}(t)$ is a particular solution of the nonhomogeneous equation (3.23).

We seek the particular solution using the method of *variation of constants* already employed in the investigation of linear differential systems. In other words, we seek $\widetilde{x}(t)$ of the form

$$\widetilde{x}(t) = c_1(t)x_1(t) + \cdots + c_n(t)x_n(t), \tag{3.29}$$

where $\{x_1, \ldots, x_n\}$ is a fundamental collection of solutions of the homogeneous equation (3.22), and $c_1, \ldots, c_n$ are unknown functions determined from the system

$$\begin{aligned} c_1' x_1 + \cdots + c_n' x_n &= 0 \\ c_1' x_1' + \cdots + c_n' x_n' &= 0 \\ \vdots \qquad & \;\;\vdots \;\; \vdots \\ c_1' x_1^{(n-1)} + \cdots + c_n' x_n^{(n-1)} &= f(t). \end{aligned} \tag{3.30}$$

The determinant of the above system is the Wronskian of the collection $\{x_1, \ldots, x_n\}$ and it is nonzero since this is a fundamental collection. Thus the above system has a unique solution. It is now easy to verify that the function $\widetilde{x}$ given by (3.29) and (3.30) is indeed a solution to (3.23).

We conclude this section with a brief discussion on the power-series method of solving higher order differential equations. For simplicity, we will limit our discussion to second-order ODEs,

$$x''(t) + p(t)x'(t) + q(t)x(t) = 0, \quad t \in I. \tag{3.31}$$

We will assume that the functions $p(t)$ and $q(t)$ are real analytic on I and so, for each t_0, $\exists R > 0$ such that

$$p(t) = \sum_{n=0}^{\infty} p_n (t - t_0)^n, \quad q(t) = \sum_{n=0}^{\infty} q_n (t - t_0)^n, \quad \forall |t - t_0| < R. \tag{3.32}$$

We seek a solution of (3.31) satisfying the Cauchy conditions

$$x(t_0) = x_0, \quad x'(t_0) = x_1, \tag{3.33}$$

described by a power series

$$x(t) = \sum_{n=0}^{\infty} \alpha_n (t - t_0)^n. \tag{3.34}$$

Equation (3.31) leads to the recurrence relations

$$(k+2)(k+1)\alpha_{k+2} + \sum_{j=0}^{k} \big((j+1)p_{k-j}\alpha_{j+1} + q_{k-j}\alpha_j \big) = 0, \quad \forall k = 0, 1, \ldots . \tag{3.35}$$

These relations together with the initial conditions (3.33) determine the coefficients α_k uniquely. One can verify directly that the resulting power series (3.34) has positive radius of convergence. Thus, the Cauchy problem (3.31) and (3.33) admits a unique *real analytic* solution.

3.5 Higher Order Linear Differential Equations with Constant Coefficients

In this section, we will deal with the problem of finding a fundamental collection of solutions for the differential equation

$$x^{(n)} + a_1 x^{(n-1)} + \cdots + a_{n-1}x' + a_n x = 0, \tag{3.36}$$

where $a_1, \ldots, a_n$ are *real* constants. The *characteristic polynomial* of the differential equation (3.36) is the algebraic polynomial

$$L(\lambda) = \lambda^n + a_1\lambda^{n-1} + \cdots + a_n. \tag{3.37}$$

To any polynomial of degree $\leq n$,

$$P(\lambda) = \sum_{k=0}^{n} p_k \lambda^k,$$

we associate the differential operator

$$P(D) = \sum_{k=0}^{n} p_k D^k, \quad D^k := \frac{d^k}{dt^k}. \tag{3.38}$$

This acts on the space $C^n(\mathbb{R})$ of functions n-times differentiable on $\mathbb{R}$ with continuous n-th order derivatives according to the rule

$$x \mapsto P(D)x := \sum_{k=0}^{n} p_k D^k x = \sum_{k=0}^{n} p_k \frac{d^k x}{dt^k}. \tag{3.39}$$

Note that (3.36) can be rewritten in the compact form

$$L(D)x = 0.$$

The key fact for the problem at hand is the following equality

$$L(D)e^{\lambda t} = L(\lambda)e^{\lambda t}, \quad \forall t \in \mathbb{R}, \quad \lambda \in \mathbb{C}. \tag{3.40}$$

From equality (3.40), it follows that, if λ is a root of the characteristic polynomial, then $e^{\lambda t}$ is a solution of (3.36). If λ is a root of multiplicity $m(\lambda)$ of $L(\lambda)$, then we define

$$S_\lambda := \begin{cases} \{e^{\lambda t}, \dots t^{m(\lambda)-1}e^{\lambda t}\}, & \text{if } \lambda \in \mathbb{R}, \\ \{\mathbf{Re}\, e^{\lambda t},\ \mathbf{Im}\, e^{\lambda t}, \dots t^{m(\lambda)-1}\, \mathbf{Re}\, e^{\lambda t},\ t^{m(\lambda)-1}\, \mathbf{Im}\, e^{\lambda t}\}, & \text{if } \lambda \in \mathbb{C} \setminus \mathbb{R}, \end{cases}$$

where **Re** and respectively **Im** denote the *real* and respectively *imaginary* part of a complex number. Note that, since the coefficients $a_1, \dots, a_n$ are real, we have

$$S_\lambda = S_{\bar{\lambda}},$$

for any root λ of L, where $\bar{\lambda}$ denotes the complex conjugate of λ. Moreover, if $\lambda = a + ib$, $b \neq 0$, is a root with multiplicity $m(\lambda)$, then

$$S_\lambda = \{e^a \cos bt, e^a \sin bt, \dots, t^{m(\lambda)-1}e^a \cos bt, t^{m(\lambda)-1}e^a \cos bt\}.$$

Theorem 3.10 *Let $\mathcal{R}$ be the set of roots of the characteristic polynomial $L(\lambda)$. For each $\lambda \in \mathcal{R}_L$ we denote by $m(\lambda)$ its multiplicity. Then the collection*

$$S := \bigcup_{\lambda \in \mathcal{R}_L} S_\lambda \tag{3.41}$$

is a fundamental collection of solutions of equation (3.36).

Proof The proof relies on the following generalization of the product formula.

Lemma 3.1 *For any $x, y \in C^n(\mathbb{R})$, we have*

$$L(D)(xy) = \sum_{\ell=0}^{n} \frac{1}{\ell!} L^{(\ell)}(D)xD^\ell y, \tag{3.42}$$

where $L^{(\ell)}(D)$ is the differential operator associated with the polynomial $L^{(\ell)}(\lambda) := \frac{d^\ell}{d\lambda^\ell}L(\lambda)$.

Proof Using the product formula, we deduce that $L(D)(xy)$ has the form

$$L(D)(xy) = \sum_{\ell=0}^{n} L_\ell(D)xD^\ell y, \tag{3.43}$$

where $L_\ell(\lambda)$ are certain polynomials of degree $\leq n - \ell$. In (3.43) we let $x = e^{\lambda t}$, $y = e^{\mu t}$ where λ, μ are arbitrary complex numbers. From (3.40) and (3.43), we obtain the equality

$$L(\lambda + \mu) = e^{-(\lambda+\mu)t}L(D)e^{(\lambda+\mu)t} = \sum_{\ell=0}^{n} L_\ell(\lambda)\mu^\ell. \tag{3.44}$$

On the other hand, Taylor's formula implies

$$L(\lambda + \mu) = \sum_{\ell=0}^{n} \frac{1}{\ell!}L^{(\ell)}(\lambda)\mu^\ell, \quad \forall \lambda, \mu \in \mathbb{C}.$$

Comparing the last equality with (3.44), we deduce that $L_\ell(\lambda) = \frac{1}{\ell!}L^{(\ell)}(\lambda)$. □

Let us now prove that any function in the collection (3.41) is indeed a solution of (3.36). Let

$$x(t) = t^r e^{\lambda t}, \quad \lambda \in \mathcal{R}_L, \quad 0 \leq r < m(\lambda).$$

Lemma 3.1 implies that

$$L(D)x = \sum_{\ell=0}^{n} \frac{1}{\ell!}L^{(\ell)}(D)e^{\lambda t}D^\ell t^r = \sum_{\ell=0}^{r} \frac{1}{\ell!}L^{(\ell)}(\lambda)e^{\lambda t}D^\ell t^r = 0.$$

If λ is a complex number, then the above equality also implies $L(D)\,\mathbf{Re}\,x = L(D)\,\mathbf{Im}\,x = 0$.

Since the complex roots of L come in conjugate pairs, we conclude that the set S in (3.41) consists of exactly n real solutions of (3.36). To prove the theorem, it suffices to show that the functions in S are linearly independent. We argue by contradiction and we assume that they are linearly dependent. Observing that for any $\lambda \in \mathcal{R}_L$ we have

$$\mathbf{Re}\,t^r e^\lambda = \frac{t^r}{2}\big(e^{\lambda t} + e^{\bar{\lambda} t}\big), \quad \mathbf{Im}\,t^r e^\lambda = \frac{t^r}{2i}\big(e^{\lambda t} - e^{\bar{\lambda} t}\big),$$

and that the roots of L come in conjugate pairs, we deduce from the assumed linear dependence of the collection S that there exists a collection of complex polynomials $P_\lambda(t)$, $\lambda \in \mathcal{R}_L$, *not all trivial*, such that

$$\sum_{\lambda \in \mathcal{R}_L} P_\lambda(t) e^{\lambda t} = 0.$$

The following elementary result shows that such nontrivial polynomials do not exist.

Lemma 3.2 *Suppose that $\mu_1, \dots, \mu_k$ are pairwise distinct complex numbers. If $P_1(t), \dots, P_k(t)$ are complex polynomials such that*

$$P_1(t)e^{\mu_1 t} + \cdots P_k(t)e^{\mu_k t} = 0, \quad \forall t \in \mathbb{R},$$

then $P_1(t) \equiv \cdots \equiv P_k(t) \equiv 0$.

Proof We argue by induction on k. The result is obviously true for $k = 1$. Assuming that the result is true for k, we prove that it is true for $k + 1$. Suppose that $\mu_0, \mu_1, \dots, \mu_k$ are pairwise distinct complex numbers and $P_0(t), P_1(t), \dots, P_k(t)$ are complex polynomials such that

$$P_0(t)e^{\mu_0 t} + P_1(t)e^{\mu_1 t} + \cdots P_k(t)e^{\mu_k t} \equiv 0. \tag{3.45}$$

Set

$$m := \max\big\{ \deg P_0, \deg P_1, \dots, \deg P_k \big\}. \tag{3.46}$$

We deduce that

$$P_0(t) + \sum_{j=1}^{k} P_k(t)e^{z_j t} \equiv 0, \quad z_j = \mu_j - \mu_0.$$

We differentiate the above equality $(m+1)$-times and, using Lemma 3.1, we deduce that

$$0 \equiv \sum_{j=1}^{k} \left(\sum_{\ell=0}^{m+1} \frac{z_j^\ell}{\ell!} P_j^{(\ell)}(t) \right) e^{z_j t} \overset{(3.46)}{=} \sum_{j=1}^{k} P_j(t + z_j) e^{z_j t}.$$

The induction assumption implies that $P_j(t + z_j) \equiv 0$, $\forall j = 1, \dots, k$. Using this fact in (3.45), we deduce that

$$P_0(t)e^{\mu_u t} \equiv 0,$$

so that $P_0(t)$ is also identically zero. □

This completes the proof of Theorem 3.10.

Let us briefly discuss the nonhomogeneous equation associated with (3.36), that is, the equation

$$x^{(n)} + a_1 x^{(n-1)} + \cdots + a_n x = f(t), \quad t \in I. \tag{3.47}$$

We have seen that the knowledge of a fundamental collection of solutions to the homogeneous equations allows us to determine a solution of the nonhomogeneous equation by using the method of variation of constants. When the equation has constant coefficients and $f(t)$ has the special form described below, this process simplifies.

A complex-valued function $f : I \to \mathbb{C}$ is called a *quasipolynomial* if it is a linear combination, with complex coefficients, of functions of the form $t^k e^{\mu t}$, where $k \in \mathbb{Z}_{\geq 0}$, $\mu \in \mathbb{C}$. A real-valued function $f : I \to \mathbb{R}$ is called a *quasipolynomial* if it is the real part of a complex polynomial. For example, the functions $t^k e^{at} \cos bt$ and $t^k e^{at} \sin bt$ are real quasipolynomials.

We want to explain how to find a complex-valued solution $x(t)$ of the differential equation

$$L(D)x = f(t),$$

where $f(t)$ is a complex quasipolynomial. Since $L(D)$ has *real* coefficients, and $x(t)$ is a solution of the above equation, we have

$$L(D)\,\mathbf{Re}\,x = \mathbf{Re}\,f(t).$$

By linearity, we can reduce the problem to the special situation when

$$f(t) = P(t)e^{\gamma t}, \tag{3.48}$$

where $P(t)$ is a complex polynomial and $\gamma \in \mathbb{C}$.

Suppose that γ is a root of order ℓ of the characteristic polynomial $L(\lambda)$. (When $\ell = 0$, this means that $L(\gamma) \neq 0$.) We seek a solution of the form

$$x(t) = t^{\ell} Q(t) e^{\gamma t}, \tag{3.49}$$

where Q is a complex polynomial to be determined. Using Lemma 3.1, we deduce from the equality $L(D)x = f(t)$ that

$$P(t) = \sum_{k=0}^{n} \frac{1}{k!} L^{(k)}(\gamma) D^k\big(t^{\ell} Q(t)\big) = \sum_{k=\ell}^{n} \frac{1}{k!} L^{(k)}(\gamma) D^k\big(t^{\ell} Q(t)\big). \tag{3.50}$$

The last equality leads to an upper triangular linear system in the coefficients of $Q(t)$ which can then be determined in terms of the coefficients of $P(t)$.

We will illustrate the above general considerations on a physical model described by a second-order linear differential equation.

3.5.1 The Harmonic Oscillator

Consider the equation of the harmonic oscillator in the presence of friction (see Sect. 1.3.4)

$$mx'' + bx' + \omega^2 x = f(t), \quad t \in \mathbb{R}, \tag{3.51}$$

where m, b, ω^2 are positive constants. The associated characteristic equation

$$m\lambda^2 + b\lambda + \omega^2 = 0$$

has roots

$$\lambda_{1,2} = -\frac{b}{2m} \pm \sqrt{\left(\frac{b}{2m}\right)^2 - \frac{\omega^2}{m}}.$$

We distinguish several cases.

1. $b^2 - 4m\omega^2 > 0$. This corresponds to the case where the friction coefficient b is "large", λ_1 and λ_2 are real, and the general solution of (3.51) has the form

   $$x(t) = C_1 e^{\lambda_1 t} + C_2 e^{\lambda_2 t} + \widetilde{x}(t),$$

   where C_1 and C_2 are arbitrary constants, and $\widetilde{x}(t)$ is a particular solution of the nonhomogeneous equation.
   The function $\widetilde{x}(t)$ is called a "*forced solution*" of the equation. Since λ_1 and λ_2 are negative, in the absence of the external force f, the motion dies down fast, converging exponentially to 0.
2. $b^2 - 4m\omega^2 = 0$. In this case,

   $$\lambda_1 = \lambda_2 = \frac{b}{2m},$$

   and the general solution of equation (3.51) has the form

   $$x(t) = C_1 e^{-\frac{bt}{2m}} + C_2 t e^{-\frac{bt}{2m}} + \widetilde{x}(t).$$

3. $b^2 - 4m\omega^2 < 0$. This is the most interesting case from a physics viewpoint. In this case,

   $$\lambda_1 = -\frac{b}{2m} + i\beta, \quad \lambda_2 = -\frac{b}{2m} - i\beta,$$

   where $\beta^2 = -\left(\frac{b}{2m}\right)^2 + \frac{\omega^2}{m}$.

   According to the general theory, the general solution of (3.51) has the form

   $$\left(C_1 \cos \beta t + C_2 \sin \beta t\right) e^{-\frac{bt}{2m}} + \widetilde{x}(t). \tag{3.52}$$

Let us assume that the external force has a harmonic character as well, that is,

$$f(t) = a\cos\nu t \quad \text{or} \quad f(t) = a\sin\nu t,$$

where the frequency ν and the amplitude a are real, nonzero, constants. We seek a particular solution of the form

$$\widetilde{x}(t) = a_1 \cos\nu t + a_2 \sin\nu t.$$

When $f = a\cos\nu t$, we find

$$\widetilde{x}(t) = \frac{a(\omega^2 - m\nu^2)\cos\nu t - ab\nu\sin\nu t}{(\omega^2 - m\nu^2)^2 + b^2\nu^2}. \tag{3.53}$$

Interestingly, as $t \to \infty$, the general solution (3.52) is asymptotic to the particular solution (3.53), that is,

$$\lim_{t\to\infty} \big(x(t) - \widetilde{x}(t) \big) = 0,$$

so that, for t sufficiently large, the general solution is practically indistinguishable from the particular forced solution $\widetilde{x}$.

Consider now the case when the frequency ν of the external perturbation is equal to the characteristic frequency of the oscillatory system, that is, $\nu = \frac{\omega}{\sqrt{m}}$. Then

$$\widetilde{x} = -\frac{a\sin\nu t}{b\nu}. \tag{3.54}$$

As can be seen from (3.54), the amplitude of $\widetilde{x}$ is then equal to the characteristic frequency of the oscillatory system and so, if $b \approx 0$, that is, the friction is practically zero, then $\frac{a}{b\nu} \approx \infty$. This is the *resonance* phenomenon often encountered in oscillatory mechanical systems. Theoretically, it manifests itself when the friction is negligible and the frequency of the external force is equal to the characteristic frequency of the system.

Equally interesting due to its practical applications is the resonance phenomenon in the differential equation (1.61) of the oscillatory electrical circuit,

$$LI'' + RI' + C^{-1}I = f(t). \tag{3.55}$$

In this case, the characteristic frequency is

$$\nu = (LC)^{-\frac{1}{2}},$$

and the resonance phenomenon appears when the source $f = U'$ is a function of the form

$$f(t) = a\cos\nu t + b\sin\nu t,$$

and the resistance R of the system is negligible. If Eq. (3.55) describes the behavior of an oscillatory electrical circuit in a receiver, $U = \int f(t)dt$ is the difference of potential between the antenna and the ground due to a certain source broadcasting electromagnetic signals with frequency ν. Then the electrical current of the form similar to (3.53)

$$I_0(t) = \frac{a(C^{-1} - L\nu^2)\cos \nu t - aR\nu \sin \nu t}{(C^{-1} - L\nu^2)^2 + R^2\nu^2}$$

develops inside the system while the other components of the general solution (3.52) are "dying down" after a sufficiently long period of time. The amplitude of the oscillation I_0 depends on the frequency ν of the broadcasting source, but also on the internal parameters L (inductance), R (resistance) and C (capacitance). To optimally select a signal from among the signals coming from several broadcast sources it is necessary to maximize the amplitude of the current I_0. This is achieved by triggering the phenomenon of resonance, that is, by choosing a capacitance C so that the internal frequency matches the broadcast frequency ν, more specifically, $C = (L\nu^2)^{-1}$. In this way, the operation of tuning-in boils down to inducing a resonance.

3.6 Linear Differential Systems with Constant Coefficients

We will investigate the differential system

$$\boldsymbol{x}' = A\boldsymbol{x}, \quad t \in \mathbb{R}, \tag{3.56}$$

where $A = (a_{ij})_{1 \le i,j \le n}$ is a *constant*, real, $n \times n$ matrix. We denote by $S_A(t)$ the fundamental matrix of (3.56) uniquely determined by the initial condition

$$S_A(0) = \mathbb{1},$$

where $\mathbb{1}$ denotes the identity matrix.

Proposition 3.1 *The family* $\{S_A(t);\ t \in \mathbb{R}\}$ *satisfies the following properties.*

(i) $S_A(t+s) = S_A(t)S_A(s)$, $\forall t, s \in \mathbb{R}$.
(ii) $S_A(0) = \mathbb{1}$.
(iii) $\lim_{t \to t_0} S_A(t)\boldsymbol{x} = S_A(t_0)\boldsymbol{x}, \quad \forall \boldsymbol{x} \in \mathbb{R}^n, \quad t_0 \in \mathbb{R}$.

Proof The group property was already established in Theorem 2.12 and follows from the uniqueness of solutions of the Cauchy problems associated with (3.56): the functions $Z(t) = S_A(t)S_A(s)$ and $Y(t) = SA(t+s)$ both satisfy (3.56) with the initial condition $Y(0) = Z(0) = S_A(s)$. Property (ii) follows from the definition, while (iii) follows from the fact that the function $t \mapsto S_A(t)\boldsymbol{x}$ is a solution of (3.56) and, in particular, it is continuous. □

Proposition 3.1 expresses the fact that the family $\{S_A(t);\ t \in \mathbb{R}\}$ is a *one-parameter group* of linear transformations of the space $\mathbb{R}^n$. Equality (iii), which can be easily seen to hold in the stronger sense of the norm of the space of $n \times n$ matrices, expresses the continuity property of the group $S_A(t)$. The map $t \to S_A(t)$ satisfies the differential equation

$$\frac{d}{dt} S_A(t)\boldsymbol{x} = AS_A(t)\boldsymbol{x}, \quad \forall t \in \mathbb{R}, \quad \forall \boldsymbol{x} \in \mathbb{R}^n,$$

and thus

$$Ax = \frac{d}{dt}\Big|_{t=0} S_A(t)\boldsymbol{x} = \lim_{t \to 0} \frac{1}{t}\big(S_A(t)\boldsymbol{x} - \boldsymbol{x}\big). \tag{3.57}$$

Equality (3.57) expresses the fact that A is the generator of the one-parameter group $S_A(t)$.

We next investigate the structure of the fundamental matrix $S_A(t)$ and the ways we can compute it. To do this, we need to digress briefly and discuss series of $n \times n$ matrices.

Matrix-valued functions. To a sequence of $n \times n$ matrices $\{A_k\}_{k \geq 0}$, we associate the formal series

$$\sum_{k=0}^{\infty} A_k$$

and we want to give a precise meaning to equalities of the form

$$A = \sum_{k=0}^{\infty} A_k. \tag{3.58}$$

We recall (see (A.5) for more details) that the norm of an $n \times n$ matrix $A = (a_{ij})_{1 \leq i,j \leq n}$ is defined by

$$\|A\| := \max_i \sum_{j=1}^{n} |a_{ij}|.$$

Definition 3.1 We say that the series of $n \times n$ matrices $\sum_{k=0}^{\infty} A_k$ converges to A, and we express this as in (3.58), if the sequence of partial sums

$$B_N = \sum_{k=0}^{N} A_k$$

converges in norm to A, that is,

$$\lim_{N \to \infty} \|B_N - A\| = 0. \tag{3.59}$$

Since the convergence (3.59) is equivalent to entry-wise convergence, it is not hard to see that Cauchy's theorem on the convergence of numerical series is also valid in the case of matrix series. In particular, we have the following convergence criterion.

Proposition 3.2 *If the matrix series* $\sum_{k=0}^{\infty} A_k$ *is majorized by a convergent numerical series, that is,*

$$\|A_k\| \leq a_k, \quad \textit{where} \quad \sum_{k=0}^{\infty} a_k < \infty,$$

then the series $\sum_{k=0}^{\infty} A_k$ *is convergent.*

Using the concept of convergent matrix series, we can define certain functions with matrices as arguments. For example, if f is a numerical analytic function of the form

$$f(\lambda) = \sum_{j=0}^{\infty} a_j \lambda^j, \quad |\lambda| < R,$$

we then define

$$f(A) := \sum_{j=0}^{\infty} a_j A^j, \quad \|A\| < R. \tag{3.60}$$

According to Proposition 3.2, the series (3.60) is convergent for $\|A\| < R$. In this fashion, we can extend to matrices the exponential, logarithm, cos, sin, etc. In particular, for any $t \in \mathbb{R}$ we can define

$$e^{tA} = \sum_{j=0}^{\infty} \frac{t^j}{j!} A^j, \tag{3.61}$$

where A is an arbitrary $n \times n$ matrix. Proposition 3.2 implies that series (3.61) converges for any t and any A.

Theorem 3.11 $e^{tA} = S_A(t)$.

Proof Since $e^{tA}\big|_{t=0} = \mathbb{1}$, it suffices to show that e^{tA} satisfies the differential equation (3.56).

Using classical analysis results, we deduce that series (3.61) can be term-by-term differentiated and we have

$$\frac{d}{dt} e^{tA} = A e^{tA}, \quad \forall t \in \mathbb{R}, \tag{3.62}$$

which proves the claim in the theorem. □

Remark 3.3 It is worth mentioning that, in the special case of the constant coefficients system (3.56), the variation of constants formula (3.21) becomes

$$\boldsymbol{x}(t) = e^{(t-t_0)A}\boldsymbol{x}_0 + \int_{t_0}^{t} e^{(t-s)A}\boldsymbol{b}(s)ds, \quad \forall t \in \mathbb{R}.$$

We can now prove a theorem on the structure of the fundamental matrix e^{tA}.

Theorem 3.12 *The (i, j)-entry of the matrix e^{tA} has the form*

$$\sum_{\ell=1}^{k} p_{\ell,i,j}(t)e^{\lambda_\ell t}, \quad i, j = 1, \dots, n, \tag{3.63}$$

where λ_ℓ are the roots of the equation $\det(\lambda\mathbb{1} - A) = 0$*, and $p_{\ell,i,j}$ is an algebraic polynomial of degree at most $m_\ell - 1$, and m_ℓ is the multiplicity of the root λ_ℓ.*

Proof Denote by $\boldsymbol{y}(t)$ a column of the matrix e^{tA} and by $P(\lambda)$ the characteristic polynomial of the matrix A, that is, $P(\lambda) = \det(\lambda\mathbb{1} - A)$. From Cayley's theorem, we deduce that $P(A) = 0$. Using the fact that $\boldsymbol{y}(t)$ satisfies (3.56), we deduce that the column $\boldsymbol{y}(t)$ satisfies the linear differential system with constant coefficients

$$P(D)\boldsymbol{y}(t) = P(A)\boldsymbol{y}(t) = 0,$$

where the differential operator $P(D)$ is defined as in (3.38). Invoking Theorem 3.10, we deduce that the components of $\boldsymbol{y}(t)$ have the form (3.63). □

We now want to give an explicit formula for computing e^{tA} in the form of an integral in the complex plane. Denote by $\lambda_1, \dots, \lambda_k$ the eigenvalues of the matrix A and by $m_1, \dots, m_k$ their (algebraic) multiplicities.

Let Γ denote a closed contour in the complex plane that surrounds the eigenvalues $\lambda_1, \dots, \lambda_k$; see Fig. 3.1.

The next theorem is the main result of this section.

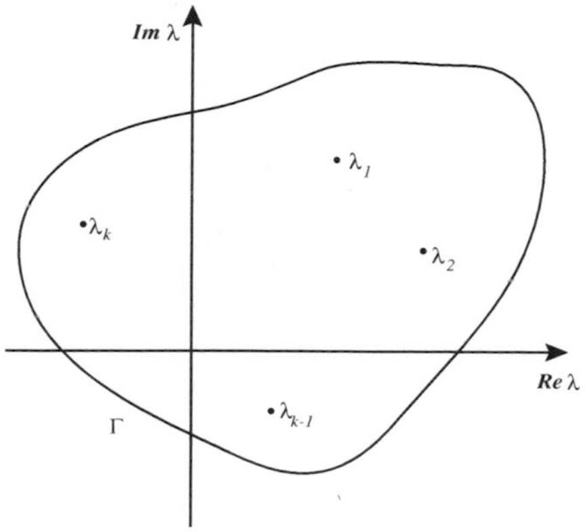

Fig. 3.1 A contour surrounding the spectrum of A

Theorem 3.13 *We have the equality*

$$e^{tA} = \frac{1}{2\pi \boldsymbol{i}} \int_{\Gamma} e^{t\lambda}(\lambda \mathbb{1} - A)^{-1} d\lambda, \quad \forall t \in \mathbb{R}. \tag{3.64}$$

Proof We denote by $X(t)$ the matrix

$$X(t) := \frac{1}{2\pi \boldsymbol{i}} \int_{\Gamma} e^{t\lambda}(\lambda \mathbb{1} - A)^{-1} d\lambda, \quad \forall t \in \mathbb{R}. \tag{3.65}$$

To prove (3.64), it suffices to verify the equalities

$$\begin{aligned} X(0) &= \mathbb{1}, \\ X'(t) &= AX(t), \quad \forall t \in \mathbb{R}. \end{aligned} \tag{3.66}$$

The equality (3.65) implies that

$$X'(t) = \frac{1}{2\pi \boldsymbol{i}} \int_{\Gamma} e^{t\lambda} \lambda(\lambda \mathbb{1} - A)^{-1} d\lambda, \quad \forall t \in \mathbb{R}.$$

Using the elementary equality

$$\lambda(\lambda \mathbb{1} - A)^{-1} = (\lambda \mathbb{1} - A)^{-1} + \mathbb{1}, \quad \forall t \in \mathbb{R},$$

we deduce that

$$X'(t) = AX(t) + \left(\frac{1}{2\pi \boldsymbol{i}} \int_{\Gamma} e^{t\lambda} d\lambda \right) \mathbb{1}.$$

From the residue formula, we deduce that

$$\int_{\Gamma} e^{t\lambda} d\lambda = 0,$$

and thus $X(t)$ satisfies the second identity in (3.66).

Let us now compute $X(0)$. Note that we have the equality

$$(\lambda \mathbb{1} - A)^{-1} = \lambda^{-1} \sum_{k=0}^{\infty} \lambda^{-k} A^k, \quad \forall |\lambda| > \|A\|. \tag{3.67}$$

To prove this equality, observe first that the series in the right-hand side is convergent for $|\lambda| > \|A\|$. Next, a simple computation shows that

$$\lambda^{-1}(\lambda \mathbb{1} - A) \sum_{k=0}^{\infty} \lambda^{-k} A^k = \mathbb{1},$$

which obviously implies (3.67).

Taking (3.67) into account, we deduce

$$X(0) = \frac{1}{2\pi \boldsymbol{i}} \int_{\Gamma} (\lambda \mathbb{1} - A)^{-1} d\lambda = \frac{1}{2\pi \boldsymbol{i}} \sum_{k=0}^{\infty} A^k \int_{\Gamma} \lambda^{-k-1} d\lambda. \tag{3.68}$$

Without loss of generality, we can assume that the contour Γ is contained inside the disk

$$\big\{ \lambda \in \mathbb{C};\ |\lambda| \leq \|A\| + \varepsilon \big\}.$$

The residue formula then implies

$$\frac{1}{2\pi \boldsymbol{i}} \int_{\Gamma} \lambda^{-k-1} d\lambda = \begin{cases} 1, & k = 0 \\ 0, & k > 0. \end{cases} \tag{3.69}$$

Using the last equality in (3.68), we deduce $X(0) = \mathbb{1}$. This completes the proof of Theorem 3.13. □

Remark 3.4 Equality (3.64) can be used to give an alternative proof of the structure theorem Theorem 3.12, and also as a method of computing e^{tA}.

We set $D(\lambda) := \det(\lambda \mathbb{1} - A)$, $A(\lambda) := \operatorname{adj}(\lambda \mathbb{1} - A) =$ *the adjoint matrix*[1] of $(\lambda \mathbb{1} - A)$. We can rewrite (3.64) in the form

$$e^{tA} = \frac{1}{2\pi \boldsymbol{i}} \int_{\Gamma} \frac{e^{t\lambda}}{D(\lambda)} A(\lambda) d\lambda = \frac{1}{2\pi \boldsymbol{i}} \int_{\Gamma} \frac{e^{t\lambda}}{(\lambda - \lambda_1)^{m_1} \cdots (\lambda - \lambda_k)^{m_k}} A(\lambda) d\lambda.$$

The residue formula then implies that

$$e^{tA} = \sum_{j=1}^{k} R(\lambda_j), \tag{3.70}$$

where we denote by $R(\lambda_j)$ the residue of the matrix-valued function

$$\lambda \mapsto \frac{e^{t\lambda}}{D(\lambda)} A(\lambda).$$

This residue can be computed using the well-known formula

$$R(\lambda_j) = \frac{1}{(m_j - 1)!} \frac{d^{m_j - 1}}{d\lambda^{m_j - 1}} \left(\frac{e^{t\lambda}(\lambda - \lambda_j)^{m_j}}{D(\lambda)} A(\lambda) \right) \Big|_{\lambda = \lambda_j}. \tag{3.71}$$

---

[1]The (i, j)-entry of $A(\lambda)$ is the (j, i)-cofactor of $(\lambda \mathbb{1} - A)$ so that $(\lambda \mathbb{1} - A)A(\lambda) = D(\lambda)\mathbb{1}$.

Equalities (3.70) and (3.71) imply Theorem 3.12. Moreover, the above process shows that the computation of e^{tA} can be performed by using algebraic operations.

3.7 Differentiability in Initial Data

In this section, we have a new look at the problem investigated in Sect. 2.6. Consider the Cauchy problem

$$\begin{aligned} \boldsymbol{x}' &= \boldsymbol{f}(t,\boldsymbol{x}), \quad (t,\boldsymbol{x}) \in \Omega \subset \mathbb{R}^{n+1}, \\ \boldsymbol{x}(t_0) &= \boldsymbol{x}_0, \end{aligned} \tag{3.72}$$

where $\boldsymbol{f} : \Omega \to \mathbb{R}^n$ is continuous in $(t,\boldsymbol{x})$ and locally Lipschitz in $\boldsymbol{x}$. We denote by $\boldsymbol{x}(t; t_0, \boldsymbol{x}_0)$ the right-saturated solution of the Cauchy problem (3.72) defined on the right-maximal existence interval $[t_0, T)$. We proved in Theorem 2.14 that for any $t' \in [t_0, T)$ there exists an $\eta > 0$ such that for any

$$\boldsymbol{\xi} \in B(\boldsymbol{x}_0, \eta) = \big\{ \boldsymbol{x} \in \mathbb{R}^n; \;\; |\boldsymbol{x} - \boldsymbol{x}_0\| \leq \eta \big\}$$

the solution $\boldsymbol{x}(t; t_0, \boldsymbol{\xi})$ is defined on $[t_0, T']$ and the resulting map

$$B(\boldsymbol{x}_0, \eta) \ni \boldsymbol{\xi} \mapsto \boldsymbol{x}(t; t_0, \boldsymbol{\xi}) \in C\big([t_0, T']; \mathbb{R}^n \big)$$

is continuous. We now investigate the differentiability of the above map.

Theorem 3.14 *Under the same assumptions as in Theorem* 2.14 *assume additionally that the function* $\boldsymbol{f}$ *is differentiable with respect to* $\boldsymbol{x}$ *and the differential* $\boldsymbol{f}_{\boldsymbol{x}}$ *is continuous with respect to* $(t,\boldsymbol{x})$. *Then the function* $\boldsymbol{x}(t; t_0, \boldsymbol{\xi})$ *is differentiable with respect to* $\boldsymbol{\xi}$ *on* $B(\boldsymbol{x}_0, \eta)$ *and its differential* $X(t) := \boldsymbol{x}_{\boldsymbol{\xi}}(t; t_0, \boldsymbol{\xi})$ *is the fundamental matrix of the linear system* (variation equation)

$$\boldsymbol{y}' = \boldsymbol{f}_{\boldsymbol{x}}\big(t, \boldsymbol{x}(t; t_0, \boldsymbol{x})\big)\boldsymbol{y}, \quad t_0 \leq t \leq T', \tag{3.73}$$

satisfying the initial condition

$$X(0) = \mathbb{1}. \tag{3.74}$$

Proof To prove that the function $\boldsymbol{x}(t; t_0, \boldsymbol{\xi})$ is differentiable in $\boldsymbol{\xi}$, we consider two arbitrary vectors $\boldsymbol{\xi}, \widetilde{\boldsymbol{\xi}} \in B(\boldsymbol{x}_0, \eta)$. We have the equality

$$\begin{aligned} \boldsymbol{x}(t; t_0, \widetilde{\boldsymbol{\xi}}) - \boldsymbol{x}(t; t_0, \boldsymbol{\xi}) - X(t)\big(\widetilde{\boldsymbol{\xi}} - \boldsymbol{\xi}\big) &= \int_{t_0}^{t} \Big(\boldsymbol{f}\big(s, \boldsymbol{x}(s; t_0, \widetilde{\boldsymbol{\xi}})\big) \\ -\boldsymbol{f}\big(s, \boldsymbol{x}(s; t_0, \boldsymbol{\xi})\big) &- \boldsymbol{f}_{\boldsymbol{x}}\big(s, \boldsymbol{x}(s; t_0, \boldsymbol{\xi})\big)X(s)\big(\widetilde{\boldsymbol{\xi}} - \boldsymbol{\xi}\big)\Big) ds, \end{aligned} \tag{3.75}$$

where $X(t)$ is the fundamental matrix of the linear system (3.73) that satisfies the initial condition (3.74). On the other hand, the mean value theorem implies that

$$\begin{aligned} & \boldsymbol{f}\big(s, \boldsymbol{x}(s; t_0, \widetilde{\boldsymbol{\xi}})\big) - \boldsymbol{f}\big(s, \boldsymbol{x}(s; t_0, \boldsymbol{\xi})\big) \\ &= \boldsymbol{f}_{\boldsymbol{x}}\big(s, \boldsymbol{x}(s; t_0, \boldsymbol{\xi})\big)\big(\boldsymbol{x}(s; t_0, \widetilde{\boldsymbol{\xi}}) - \boldsymbol{x}(s; t_0, \boldsymbol{\xi})\big) + R\big(s, \widetilde{\boldsymbol{\xi}}, \boldsymbol{\xi}\big). \end{aligned} \tag{3.76}$$

Since f is locally Lipschitz, there exists a constant $L > 0$ such that, $\forall t \in [t_0, T']$, we have

$$\|\boldsymbol{x}(t; t_0, \widetilde{\boldsymbol{\xi}}) - \boldsymbol{x}(t; t_0, \boldsymbol{\xi})\| \leq \|\widetilde{\boldsymbol{\xi}} - \boldsymbol{\xi}\| + L \int_{t_0}^{t} \|\boldsymbol{x}(s; t_0, \widetilde{\boldsymbol{\xi}}) - \boldsymbol{x}(s; t_0, \boldsymbol{\xi})\| ds.$$

Invoking Gronwall's lemma, we deduce that

$$\|\boldsymbol{x}(s; t_0, \widetilde{\boldsymbol{\xi}}) - \boldsymbol{x}(t; t_0, \boldsymbol{\xi})\| \leq \|\widetilde{\boldsymbol{\xi}} - \boldsymbol{\xi}\| e^{L(s-t_0)}, \quad \forall s \in [t_0, T']. \tag{3.77}$$

Inequality (3.77) and the continuity of the differential $\boldsymbol{f}_{\boldsymbol{x}}$ imply that the remainder R in (3.76) satisfies the estimate

$$\|R\big(s, \widetilde{\boldsymbol{\xi}}, \boldsymbol{\xi}\big)\| \leq \omega(\widetilde{\boldsymbol{\xi}}, \boldsymbol{\xi}) \|\widetilde{\boldsymbol{\xi}} - \boldsymbol{\xi}\|, \tag{3.78}$$

where

$$\lim_{\|\widetilde{\boldsymbol{\xi}} - \boldsymbol{\xi}\| \to 0} \omega(\widetilde{\boldsymbol{\xi}}, \boldsymbol{\xi}) = 0.$$

Using (3.76) in (3.75), we obtain the estimate

$$z(t) \leq (T' - t_0)\omega(\widetilde{\boldsymbol{\xi}}, \boldsymbol{\xi}) \|\widetilde{\boldsymbol{\xi}} - \boldsymbol{\xi}\| + L_1 \int_{t_0}^{t} z(s) ds, \quad \forall t \in [t_0, T'], \tag{3.79}$$

where

$$z(t) := \big\| \boldsymbol{x}(t; t_0, \widetilde{\boldsymbol{\xi}}) - \boldsymbol{x}(t; t_0, \boldsymbol{\xi}) - X(t)\big(\widetilde{\boldsymbol{\xi}} - \boldsymbol{\xi}\big) \big\|.$$

Invoking Gronwall's lemma again, we deduce that

$$\begin{aligned} & \big\| \boldsymbol{x}(t; t_0, \widetilde{\boldsymbol{\xi}}) - \boldsymbol{x}(t; t_0, \boldsymbol{\xi}) - X(t)\big(\widetilde{\boldsymbol{\xi}} - \boldsymbol{\xi}\big) \big\| \\ & \leq (T' - t_0)\omega(\widetilde{\boldsymbol{\xi}}, \boldsymbol{\xi}) \|\widetilde{\boldsymbol{\xi}} - \boldsymbol{\xi}\| e^{L_1(T' - t_0)} = o\big(\|\widetilde{\boldsymbol{\xi}} - \boldsymbol{\xi}\|\big). \end{aligned} \tag{3.80}$$

The last inequality implies (see Appendix A.5) that

$$\boldsymbol{x}_{\boldsymbol{\xi}}(t; t_0, \boldsymbol{\xi}) = X(t). \qquad \square$$

We next investigate the differentiability with respect to a parameter $\boldsymbol{\lambda}$ of the solution $\boldsymbol{x}(t; t_0, \boldsymbol{x}_0, \lambda)$ of the Cauchy problem

$$\boldsymbol{x}' = \boldsymbol{f}(t, \boldsymbol{x}, \boldsymbol{\lambda}), \quad (t, \boldsymbol{x}) \in \Omega \subset \mathbb{R}^{n+1}, \quad \boldsymbol{\lambda} \in U \subset \mathbb{R}^m, \tag{3.81}$$

$$\boldsymbol{x}(t_0) = \boldsymbol{x}_0. \tag{3.82}$$

The parameter $\boldsymbol{\lambda} = (\lambda_1, \ldots, \lambda_m)$ varies in a bounded open subset U on $\mathbb{R}^m$. Fix $\boldsymbol{\lambda}_0 \in U$. Assume that the right-saturated solution $\boldsymbol{x}(t; t_0, \boldsymbol{x}_0, \boldsymbol{\lambda}_0)$ of (3.81) and (3.82) corresponding to $\boldsymbol{\lambda} = \boldsymbol{\lambda}_0$ is defined on the maximal-to-the-right interval $[t_0, T)$. We have

Theorem 3.15 *Let*

$$f : \Omega \times U \to \mathbb{R}^n$$

be a continuous function, differentiable in the $\boldsymbol{x}$ *and* $\boldsymbol{\lambda}$ *variables, and with the differentials* $\boldsymbol{f}_{\boldsymbol{x}}, \boldsymbol{f}_{\boldsymbol{\lambda}}$ *continuous in* $(t, \boldsymbol{x}, \boldsymbol{\lambda})$. *Then, for any* $T' \in [t_0, T)$, *there exists a* $\delta > 0$ *such that the following hold.*

(i) The solution $\boldsymbol{x}(t; t_0, \boldsymbol{x}_0, \boldsymbol{\lambda})$ *is defined on* $[t_0, T']$ *for any*

$$\boldsymbol{\lambda} \in B(\boldsymbol{\lambda}_0, \delta) := \left\{ \boldsymbol{\lambda} \in \mathbb{R}^m; \quad \|\boldsymbol{\lambda} - \boldsymbol{\lambda}_0\| \le \delta \right\}.$$

(ii) For any $t \in [t_0, T']$, *the map* $B(\boldsymbol{\lambda}_0, \delta) \ni \boldsymbol{\lambda} \mapsto \boldsymbol{x}(t; t_0, \boldsymbol{x}_0, \boldsymbol{\lambda}) \in \mathbb{R}^n$ *is differentiable and the differential* $\boldsymbol{y}(t) := \boldsymbol{x}_{\boldsymbol{\lambda}}(t; t_0, \boldsymbol{x}_0, \boldsymbol{\lambda}) : [t_0, T'] \to \mathbb{R}^n \times \mathbb{R}^m$ *is uniquely determined by the (matrix-valued) linear Cauchy problem*

$$\begin{aligned} \boldsymbol{y}'(t) = \boldsymbol{f}_{\boldsymbol{x}}\big(t, \boldsymbol{x}(t; t_0, \boldsymbol{x}_0, \boldsymbol{\lambda}), \boldsymbol{\lambda}\big)\boldsymbol{y}(t) + \boldsymbol{f}_{\boldsymbol{\lambda}}\big(t, \boldsymbol{x}(t; t_0, \boldsymbol{x}_0, \boldsymbol{\lambda}), \boldsymbol{\lambda}\big), \\ \forall t \in [t_0, T'], \end{aligned} \tag{3.83}$$

$$\boldsymbol{y}(t_0) = 0. \tag{3.84}$$

Proof As in the proof of Theorem 2.15, we can write (3.81) and (3.82) as a new Cauchy problem,

$$\begin{aligned} \boldsymbol{x}' &= \boldsymbol{F}(t, z), \\ z(t_0) &= \boldsymbol{\zeta} := (\boldsymbol{\xi}, \boldsymbol{\lambda}), \end{aligned} \tag{3.85}$$

where $z = (\boldsymbol{x}, \boldsymbol{\lambda})$, $\boldsymbol{F}(t, z) = \big(\boldsymbol{f}(t, \boldsymbol{x}, \boldsymbol{\lambda}), 0\big) \in \mathbb{R}^n \times \mathbb{R}^m$. According to Theorem 3.14, the map

$$\boldsymbol{\zeta} \mapsto \big(\boldsymbol{x}(t; t_0, \boldsymbol{\xi}, \lambda), \lambda_0\big) =: z(t; t_0, \boldsymbol{\zeta})$$

is differentiable and its differential

$$Z(t) := \frac{\partial z}{\partial \boldsymbol{\zeta}} = \begin{bmatrix} \frac{\partial \boldsymbol{x}}{\partial \boldsymbol{\xi}}(t; t_0, \boldsymbol{\xi}, \boldsymbol{\lambda}) & \frac{\partial \boldsymbol{x}}{\partial \boldsymbol{\lambda}}(t; t_0, \boldsymbol{\xi}, \boldsymbol{\lambda}) \\ 0 & \mathbb{1}_m \end{bmatrix}$$

($\mathbb{1}_m$ is the identity $m \times m$ matrix) satisfies the differential equation

$$Z'(t) = \boldsymbol{F}_z(t, z)Z(t), \quad Z(t_0) = \mathbb{1}_{n+m}. \tag{3.86}$$

Taking into account the description of $Z(t)$ and the equality

$$\boldsymbol{F}_z(t,z) = \begin{bmatrix} \boldsymbol{f}(t,\boldsymbol{x},\boldsymbol{\lambda})\, \boldsymbol{f}_{\boldsymbol{\lambda}}(t,\boldsymbol{x},\boldsymbol{\lambda}) \\ 0 \quad\quad 0 \end{bmatrix},$$

we conclude from (3.86) that $\boldsymbol{y}(t) := \boldsymbol{x}_{\boldsymbol{\lambda}}(t;\, t_0, \boldsymbol{x}_0, \boldsymbol{\lambda})$ satisfies the Cauchy problem (3.83) and (3.84). □

Remark 3.5 The matrix $\boldsymbol{x}_{\boldsymbol{\lambda}}(t;\, t_0, \boldsymbol{x}_0, \boldsymbol{\lambda})$ is sometimes called the *sensitivity matrix* and its entries are known as *sensitivity functions*. Measuring the changes in the solution under small variations of the parameter $\boldsymbol{\lambda}$, this matrix is an indicator of the robustness of the system.

Theorem 3.15 is especially useful in the approximation of solutions of differential systems via the so-called *small-parameter method*.

Let us denote by $\boldsymbol{x}(t, \boldsymbol{\lambda})$ the solution $\boldsymbol{x}(t;\, t_0, \boldsymbol{x}_0, \boldsymbol{\lambda})$ of the Cauchy problem (3.81) and (3.82). We then have a first-order approximation

$$\boldsymbol{x}(t,\boldsymbol{\lambda}) = \boldsymbol{x}(t,\boldsymbol{\lambda}_0) + \boldsymbol{x}_{\boldsymbol{\lambda}}(t,\boldsymbol{\lambda}_0)(\boldsymbol{\lambda}-\boldsymbol{\lambda}_0) + o(\|\boldsymbol{\lambda}-\boldsymbol{\lambda}_0\|), \tag{3.87}$$

where $\boldsymbol{y}(t) = \boldsymbol{x}_{\boldsymbol{\lambda}}(t, \boldsymbol{\lambda}_0)$ is the solution of the variation equation (3.83) and (3.84). Thus, in a neighborhood of the parameter $\boldsymbol{\lambda}_0$, we have

$$\boldsymbol{x}(t,\boldsymbol{\lambda}) \approx \boldsymbol{x}(t,\boldsymbol{\lambda}_0) + \boldsymbol{x}_{\boldsymbol{\lambda}}(t,\boldsymbol{\lambda}_0)(\boldsymbol{\lambda}-\boldsymbol{\lambda}_0).$$

We have thus reduced the approximation problem to solving a linear differential system. Let us illustrate the technique on the following example

$$x' = x + \lambda t x^2 + 1, \quad x(0) = 1, \tag{3.88}$$

where λ is a sufficiently small parameter. Equation (3.88) is a Riccati type equation and cannot be solved explicitly. However, for $\lambda = 0$ it reduces to a linear equation and its solution is

$$x(t,0) = 2e^t - 1.$$

According to formula (3.87), the solution $x(t, \lambda)$ admits an approximation

$$x(t,\lambda) = 2e^t - 1 + \lambda y(t) + o(|\lambda|),$$

where $y(t) = x_\lambda(t, 0)$ is the solution of the variation equation

$$y' = y + t(2e^t - 1)^2, \quad y(0) = 0.$$

Hence

$$y(t) = \int_0^t s(2e^s - 1)^2 e^{t-s} ds.$$

Thus, for small values of the parameter λ, the solution to problem (3.88) is well approximated by

$$2e^t - 1 + \lambda e^t\big(4te^t - 2t^2 - 4e^t + e^{-t} + 3 - te^{-t}\big).$$

3.8 Distribution Solutions of Linear Differential Equations

The notion of distribution is a relatively recent extension of the concept of function (introduced by *L. Schwarz* (1915–2002)) that leads to a rethinking of the foundations of mathematical analysis. Despite its rather abstract character, the concept of distribution has, as will be seen shortly, a solid physical support, which was behind its development.

Given an open interval I of the real axis $\mathbb{R}$ (in particular, I could be the whole axis), we denote by $C^\infty(I)$ the space of infinitely differentiable (or smooth) functions on I. For any $\varphi \in C^\infty(\varphi)$, we denote by $\operatorname{supp}\varphi$ its *support*

$$\operatorname{supp}\varphi = \overline{\{t \in I;\ \ \varphi(t) \neq 0\}},$$

where $\overline{D}$ denotes the closure in I of the subset $D \subset I$. We denote by $C_0^\infty(I)$ the set of smooth functions $\varphi \in C^\infty(I)$ such that $\operatorname{supp}\varphi$ is a *compact* subset of I. It is not hard to see that $C_0^\infty(I) \neq \emptyset$. Indeed, if $I = (a, b)$ and $t_0 \in I$, then the function

$$\varphi(t) = \begin{cases} \exp\left(\frac{1}{|t-t_0|^2-\varepsilon^2}\right), & |t - t_0| < \varepsilon, \\ 0, & |t - t_0| \geq \varepsilon, \end{cases} \qquad 0 < \varepsilon < \min\{t_0 - a, b - t_0\},$$

belongs to $C^\infty(I)$.

The set $C_0^\infty(I)$ is obviously a vector space over $\mathbb{R}$ (or over $\mathbb{C}$ if we are dealing with complex-valued functions) with respect to the usual addition and scalar multiplication operations. This space can be structured as a topological space by equipping it with a notion of convergence.

Given a sequence $\{\varphi_n\}_{n\geq 1} \subset C_0^\infty(I)$, we say that it *converges* to $\varphi \in C_0^\infty(I)$, and we denote this by $\varphi_n \Rightarrow \varphi$, if there exists a compact subset $K \subset I$ such that

$$\operatorname{supp}\varphi_n \subset K, \quad \forall n, \tag{3.89}$$

$$\frac{d^j\varphi_n}{dt^j}(t) \to \frac{d^j\varphi}{dt^j}(t) \ \text{ uniformly on } K, \forall j, \tag{3.90}$$

as $n \to \infty$.

We define a *distribution* (or *generalized function*) on I to be a linear and continuous functional on $C_0^\infty(I)$, that is, a linear map $u : C_0^\infty(I) \to \mathbb{R}$ such that

$$\lim_{n\to\infty} u(\varphi_n) = u(\varphi)$$

for any sequence $\{\varphi_n\}$ in $C_0^\infty(I)$ such that $\varphi_n \Rightarrow \varphi$. We denote by $u(\varphi)$ the value of this functional at φ. The set of distributions, which is itself a vector space, is denoted by $\mathcal{D}'(I)$. We stop the flow of definitions to discuss several important examples of distributions.

Example 3.1

(a) Any function $f : I \to \mathbb{R}$ which is Riemann or Lebesgue integrable on any compact subinterval of I canonically determines a distribution $u_f : C_0^\infty(I) \to \mathbb{R}$ defined by

$$u_f(\varphi) := \int_I f(t)\varphi(t)dt, \quad \forall \varphi \in C_0^\infty(I). \tag{3.91}$$

It is not difficult to verify that u_f is a linear functional and

$$u_f(\varphi_n) \to u_f(\varphi),$$

for any sequence $\varphi_n \Rightarrow \varphi$.
On the other hand, if $u_f(\varphi) = 0$, for any $\varphi \in C_0^\infty(I)$, then obviously $f(t) = 0$ almost everywhere (a.e.) on I. Therefore, the correspondence $f \mapsto u_f$ is an injection of the space of locally integrable functions into the space of distributions on I. In this fashion, we can identify any locally integrable function on I with a distribution.

(b) *The Dirac distribution*. Let $t_0 \in \mathbb{R}$. We define the distribution δ_{t_0}

$$\delta_{t_0}(\varphi) = \varphi(t_0), \quad \forall \varphi \in C_0^\infty(\mathbb{R}). \tag{3.92}$$

It is easy to see that δ_{t_0} is indeed a linear and continuous functional on $C_0^\infty(\mathbb{R})$, that is, $\delta_{t_0} \in \mathcal{D}'(\mathbb{R})$. This is called the *Dirac distribution* (concentrated) at t_0. Historically speaking, the distribution δ_{t_0} is the first nontrivial example of a distribution and was introduced in 1930 by the physicist *P.A.M. Dirac* (1902–1984). In Dirac's interpretation, δ_{t_0} had to be a "function" on $\mathbb{R}$ which is zero everywhere but at t_0, and whose integral had to be 1. Obviously, from a physical point of view, such a "function" had to represent an impulsive force, but it did not fit into the traditional setup of mathematical analysis. We could view such a "function" as the limit as $\varepsilon \searrow 0$ of the family of functions defined by (see Fig. 3.2)

$$\eta_\varepsilon(t) := \frac{1}{\varepsilon}\eta\left(\frac{t-t_0}{\varepsilon}\right) \tag{3.93}$$

where $\eta \in C_0^\infty(\mathbb{R})$ is a function such that

$$\operatorname{supp}\eta \subset (-1,1), \quad \int_{\mathbb{R}} \eta(t)dt = 1.$$

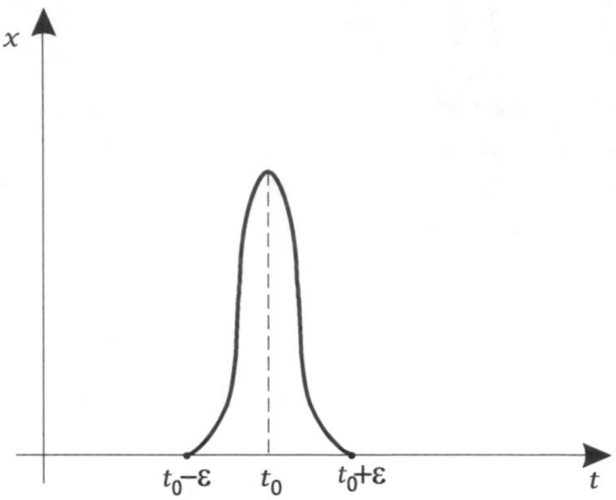

Fig. 3.2 Approximating Dirac's δ-function

Then

$$\operatorname{supp}\eta_\varepsilon \subset (t_0-\varepsilon, t_0+\varepsilon), \quad \int_{\mathbb{R}} \eta_\varepsilon(t)dt = 1,$$

and we can regard η_ε as approximating the "function" δ_{t_0}. We cannot expect that η_ε converges in any conventional way to a function that we could consider to be δ_{t_0}. However, we have the equality

$$\lim_{\varepsilon\searrow 0}\int_{\mathbb{R}} \eta_\varepsilon(t)\varphi(t)dt = \varphi(t_0) = \delta_{t_0}(\varphi), \quad \forall\varphi\in C_0^\infty(\mathbb{R}).$$

We say that the family $\{\eta_\varepsilon\}$ converges in the sense of distributions to δ_{t_0}. It is in this sense that we can restore Dirac's original intuition.

An important property of distributions is that they can be differentiated as many times as we please. For any positive integer j, we define the j-th order distribution of the distribution $u\in\mathcal{D}'(I)$ to be the distribution $u^{(j)}$ given by

$$u^{(j)}(\varphi) := (-1)^j u\big(\varphi^{(j)}\big), \quad \forall\varphi\in C_0^\infty(I). \tag{3.94}$$

Based on the definition, it is not difficult to verify that the linear functional

$$u^{(j)} : C_0^\infty(I)\to\mathbb{R}$$

is indeed continuous and thus defines a distribution on I.

If u is a distribution associated with a C^1-function f, that is,

$$u(\varphi) = u_f(\varphi) = \int_I f(t)\varphi(t)dt, \quad \forall\varphi\in C_0^\infty(I),$$

then its distributional derivative is the distribution determined by the derivative of f, that is,

$$u_f'(\varphi) = -\int_I f(t)\varphi'(t)dt = \int_I f'(t)\varphi(t) = u_{f'}(\varphi), \quad \forall \varphi \in C_0^\infty(I).$$

In general, if f is a locally integrable function, its distributional derivative is no longer a function, but the distribution u_f' defined by

$$u_f'(\varphi) = -\int_I f(t)\varphi'(t)dt, \quad \forall \varphi \in C_0^\infty(I).$$

Example 3.2 (Heaviside function) Consider the *Heaviside function* $H : \mathbb{R} \to \mathbb{R}$,

$$H(t) = \begin{cases} 1, & t \geq 0, \\ 0, & t < 0. \end{cases}$$

Its distributional derivative, which we continue to denote by H', is thus defined by

$$H'(\varphi) = -\int_{\mathbb{R}} H(t)\varphi'(t)dt = -\int_0^\infty \varphi'(t)dt = \varphi(0), \quad \forall \varphi \in C_0^\infty(\mathbb{R}).$$

Hence $H' = \delta_0$.

Given a smooth function $a \in C^\infty(I)$ and a distribution $u \in \mathcal{D}'(I)$, we can define the product $au \in \mathcal{D}'(I)$ by the equality

$$(au)(\varphi) := u(a\varphi), \quad \forall \varphi \in C_0^\infty(I).$$

We denote by $\mathcal{D}'(I; \mathbb{R}^n)$ the space of vector-valued distributions

$$\boldsymbol{y} = \begin{bmatrix} y_1 \\ \vdots \\ y_n \end{bmatrix}$$

where $y_i \in \mathcal{D}'(I)$, $\forall i = 1, \dots, n$. We set

$$\boldsymbol{y}' := \begin{bmatrix} y_1' \\ \vdots \\ y_n' \end{bmatrix}.$$

Consider the differential equation with constant coefficients

$$x^{(n)} + a_1 x^{(n-1)} + \cdots + a_n x = f \quad \text{on } I, \tag{3.95}$$

where $f \in \mathcal{D}'(I)$. A solution of (3.95) is by definition a distribution $x \in \mathcal{D}'(I)$ that satisfies (3.95) in the sense of distributions, that is,

$$x\big((-1)^n\varphi^{(n)} + (-1)^{n-1}a_1\varphi^{(n-1)} + \cdots + a_n\varphi\big) = f(\varphi), \quad \forall \varphi \in C_0^\infty(I).$$

More generally, given the differential system

$$\boldsymbol{y}' = A\boldsymbol{y} + \boldsymbol{f}, \tag{3.96}$$

where $\boldsymbol{f} \in \mathcal{D}'(I;\mathbb{R}^n)$ and A is an $n \times n$ real matrix, we say that $\boldsymbol{y} \in \mathcal{D}'(I;\mathbb{R}^n)$ is a solution of (3.96) if for any $\varphi \in C_0^\infty(I)$ we have

$$-y'(\varphi') = \sum_{j=1}^{n} y_j(a_{ij}\varphi) + f_i(\varphi), \quad \forall i = 1,\ldots,n.$$

Clearly, $x \in \mathcal{D}'(I)$ is a solution of (3.95) if and only if

$$\boldsymbol{y} = \begin{bmatrix} x \\ x' \\ \vdots \\ x^{(n-1)} \end{bmatrix}$$

is a solution of system (3.24).

If f is continuous, then, as shown in Sect. 3.5, the set of C^n-solutions (let's call them classical) of Eq. (3.95) is given by

$$C_1x_1 + \cdots + C_nx_n + \widetilde{x}, \tag{3.97}$$

where $\{x_1,\ldots,x_n\}$ is a fundamental collection of solutions of the homogeneous equation, and $\widetilde{x}$ is a particular solution of the nonhomogeneous equation. It is then natural to ask if there are distributions not contained in the family (3.97). The answer turns out to be negative and is the object of Theorem 3.16.

Theorem 3.16 *If f is a continuous function on I, then the only solutions of equation* (3.95) (*respectively system* (3.96)) *are the classical ones.*

Proof We show first that the only distribution solutions of the equation

$$x' = 0 \tag{3.98}$$

are the classical ones, that is, the constant functions.

Indeed, if $x \in \mathcal{D}'(\mathbb{R})$ is a solution of (3.98), then, by definition, we have

$$x(\varphi') = 0, \quad \forall \varphi \in C_0^\infty(\mathbb{R}).$$

Hence

$$x(\psi) = 0 \tag{3.99}$$

for any function $\psi \in C_0^\infty(\mathbb{R})$ such that $\psi = \varphi'$, $\varphi \in C_0^\infty(\mathbb{R})$, that is,

$$\int_{\mathbb{R}} \psi(t)dt = 0.$$

Fix an arbitrary function $\chi \in C_0^\infty(\mathbb{R})$ such that

$$\int_{\mathbb{R}} \chi(t)dt = 0.$$

For any $\varphi \in C_0^\infty(\mathbb{R})$, we have

$$\varphi(t) = \chi(t) \int_{\mathbb{R}} \varphi(s)ds + \varphi(t) - \chi(t) \int_{\mathbb{R}} \varphi(s)ds,$$

and thus, according to (3.99),

$$x(\varphi) = x(\chi) \int_{\mathbb{R}} \varphi(t)dt + x\left(\varphi - \chi \int_{\mathbb{R}} \varphi(t)dt\right) = x(\chi) \int_{\mathbb{R}} \varphi(t)dt.$$

If we set $C := x(\chi)$, we deduce that

$$x(\varphi) = \int C\varphi(t)dt, \quad \forall \varphi \in C_0^\infty(\mathbb{R}),$$

and thus $x = C$.

Consider now the equation

$$x' = g \in C(I). \tag{3.100}$$

This equation can also be rewritten in the form

$$(x - G)' = 0, \quad G(t) := \int_{t_0}^{t} g(s)ds, \quad t_0 \in I.$$

From the previous discussion, we infer that $x = G + C$, where C is a constant.

If $\boldsymbol{g} \in C(I; \mathbb{R}^n)$ and $\boldsymbol{y} \in \mathcal{D}'(I; \mathbb{R}^n)$ satisfies $\boldsymbol{y}' = \boldsymbol{g}$, then, as above, we conclude that

$$(\boldsymbol{y} - \boldsymbol{G})' = 0,$$

where

$$\boldsymbol{G}(t) = \int_{t_0}^{T} \boldsymbol{g}(s)ds, \quad t_0 \in I,$$

and we infer as before that

$$y = G + c,$$

with c a constant vector in $\mathbb{R}^n$.

Consider now a solution $y \in \mathcal{D}'(I; \mathbb{R}^n)$ of system (3.96), where $f \in C(I; \mathbb{R}^n)$. If we define the product $e^{-tA}y \in \mathcal{D}'(I; \mathbb{R}^n)$ as above, we observe that we have the equality of distributions

$$\big(e^{-tA}y\big)' = (e^{-tA})'y + e^{-tA}by',$$

and thus

$$\big(e^{-tA}y\big)' = e^{-tA}f \text{ in } \mathcal{D}'(I; \mathbb{R}^n).$$

The above discussion shows that

$$e^{-tA}y = c + \int_{t_0}^{t} e^{-sA}f(s)ds \text{ on } I.$$

Thus, the distribution solutions of (3.96) are the classical solutions given by the formula of variation of constants (3.21). In particular, it follows that the only distribution solutions of (3.95) are the classical ones, given by (3.97). □

We close this section with some examples.

Example 3.3 Consider the differential equation

$$x' + ax = \mu\delta_0, \text{ on } \mathbb{R}, \tag{3.101}$$

where δ_0 denotes the Dirac distribution concentrated at the origin and $\mu \in \mathbb{R}$. If $x \in \mathcal{D}'(\mathbb{R})$ is a solution of (3.101), then, by definition,

$$-x(\varphi') + ax(\varphi) = \mu\varphi(0), \quad \forall \varphi \in C_0^\infty(\mathbb{R}). \tag{3.102}$$

If, in (3.102), we first choose φ such that $\operatorname{supp}\varphi \subset (-\infty, 0)$, and $\operatorname{supp}\varphi \subset (0, \infty)$, then we deduce that

$$x' + ax = 0 \text{ on } \mathbb{R} \setminus \{0\}$$

in the sense of distributions. Thus, according to Theorem 3.16, x is a classical solution to $x' + ax = 0$ on $\mathbb{R} \setminus \{0\}$. Hence, there exist constants $C_1, C_2 \in \mathbb{R}$ such that

$$x(t) = \begin{cases} C_1^{-at}, & t > 0, \\ C_2 e^{-at}, & t < 0. \end{cases} \tag{3.103}$$

On the other hand, the function x defined by (3.103) is locally integrable on $\mathbb{R}$ and of class C^1 on $\mathbb{R} \setminus \{0\}$. We have, therefore,

$$\begin{aligned} x(\varphi') = \int_{\mathbb{R}} x(t)\varphi'(t)dt &= \int_{-\infty}^{0} x(t)\varphi'(t)dt + \int_{0}^{\infty} x(t)\varphi'(t)dt \\ &= -\int_{-\infty}^{0} \dot{x}(t)\varphi(t)dt + x(0^-)\varphi(0) \\ &\quad - x(0^+)\varphi(0) - \int_{0}^{\infty} \dot{x}(t)\varphi(t)dt \\ &= -\int_{\mathbb{R}} \dot{x}(t)\varphi(t)dt + \varphi(0)\big(x(0^-) - x(0^+)\big), \end{aligned} \tag{3.104}$$

where $\dot{x}$ denotes the *usual* derivative of x on $\mathbb{R} \setminus \{0\}$. We rewrite (3.102) in the form

$$x(\varphi') = a \int_{\mathbb{R}} x(t)\varphi(t)dt - \mu\varphi(0), \quad \forall \varphi \in C_0^\infty(\mathbb{R}). \tag{3.105}$$

Using (3.104) in the above equality, and taking into account that $\dot{x} = -ax$ on $\mathbb{R} \setminus \{0\}$, we deduce that

$$\varphi(0)\big(x(0^-) - x(0^+)\big) = -\mu\varphi(0), \quad \forall \varphi \in C_0^\infty(\mathbb{R}).$$

Hence

$$x(0^+) - x(0^-) = \mu,$$

and thus

$$x(t) = \begin{cases} C^{-at}, & t > 0, \\ (C - \mu)e^{-at}, & t < 0, \end{cases}$$

where C is a constant. As was to be expected, the solutions of (3.101) are not of class C^1, they are not even continuous. They have a jump of size μ at $t = 0$.

Example 3.4 Consider the second-order differential equation

$$mx'' + bx' + \omega^2 x = \mu\delta_0. \tag{3.106}$$

Taking into account the interpretation of the Dirac distribution δ_0, Eq. (3.106) describes an elastic motion of a particle of mass m that was acted upon by an impulsive force of size μ at time $t = 0$.

Such phenomena appear, for example, in collisions or in cases when a cord or an elastic membrane is hit. A physical analysis of this phenomenon leads us to the conclusion that the effect of the impulsive force $\mu\delta_0$ is to produce a jump in the velocity, that is, an instantaneous change in the momentum of the particle. We will reach a similar conclusion theoretically, by solving Eq. (3.106).

If the distribution x is a solution of (3.106), then

$$\int_{\mathbb{R}} x(t)\big(m\varphi''(t) - b\varphi'(t) + \omega^2\varphi(t)\big)dt = \varphi(0), \quad \forall \varphi \in C_0^\infty(\mathbb{R}). \tag{3.107}$$

Choosing φ such that supp $\varphi(-\infty, 0)$ and then supp$(0, \infty)$, we deduce that x satisfies, in the sense of distributions, the equation

$$mx'' + bx' + \omega^2 x = \quad \text{on} \ \ \mathbb{R} \setminus \{0\}.$$

According to Theorem 3.16, the distribution x is a classical solution of this equation on each of the intervals $(-\infty, 0)$ and $(0, \infty)$. Integrating by parts, we deduce that

$$\begin{aligned} \int_{\mathbb{R}} x(t)\varphi''(t)dt &= \int_{-\infty}^{0} x(t)\varphi''(t)dt + \int_{0}^{\infty} x(t)\varphi''(t)dt \\ &= \big(x(0^-) - x(0^+)\big)\varphi'(0) + \big(\dot{x}(0^+) - \dot{x}(0^-)\big)\varphi(0) \\ &+ \int_{-\infty}^{0} \ddot{x}(t)\varphi(t)dt + \int_{0}^{\infty} \ddot{x}(t)\varphi(t)dt, \end{aligned}$$

and

$$\int_{\mathbb{R}} x(t)\varphi'(t)dt = \big(x(0^-) - x(0^+)\big)\varphi(0) = \int_{-\infty}^{0} \dot{x}(t)\varphi(t)dt - \int_{0}^{\infty} \dot{x}(t)\varphi(t)dt,$$

where $\dot{x}$ and $\ddot{x}$ denote, respectively, the first and the second classical derivative of x on $\mathbb{R} \setminus \{0\}$. Taking into account the equalities

$$m\dot{x} + b\dot{x} + \omega^2 x = 0 \ \ \text{on} \ \ \mathbb{R} \setminus \{0\}$$

and (3.107), we deduce that

$$\begin{aligned} &m\big(x(0^-) - x(0^+)\big)\varphi'(0) + m\big(\dot{x}(0^+) - \dot{x}(0^-)\big)\varphi(0) \\ &+ b\big(x(0^-) - x(0^+)\big)\varphi(0) = \mu\varphi(0), \ \ \forall \varphi \in C_0^\infty(\mathbb{R}). \end{aligned}$$

Hence

$$x(0^-) = x(0^+), m \ \ \dot{x}(0^+) - \dot{x}(0^-) = \frac{\mu}{m}.$$

Thus, the function x is continuous, but its derivative $\dot{x}$ is discontinuous at the origin where it has a jump of size $\frac{\mu}{m}$. In other words, x is the usual solution of the following system

$$\begin{cases} m\ddot{x}(t) + b\dot{x}(t) + \omega^2 x(t) = 0, & t < 0, \\ x(0) = x_0, \ \ \dot{x}(0) = x_1, \\ m\ddot{x}(t) + b\dot{x}(t) + \omega^2 x(t) = 0, & t > 0, \\ x(0) = x_0, \ \ \dot{x}(0) = x_1 + \dfrac{\mu}{m}. \end{cases}$$

Problems

3.1 Consider the differential system

$$\boldsymbol{x}' = A(t)\boldsymbol{x}, \quad t \geq 0, \tag{3.108}$$

where $A(t)$ is an $n \times n$ matrix whose entries depend continuously on $t \in [0, \infty)$ and satisfies the condition

$$\liminf_{t \to \infty} \int_0^t \operatorname{tr} A(s)ds > -\infty. \tag{3.109}$$

Let $X(t)$ be a fundamental matrix of system (3.108) that is bounded as a function of $t \in [0, \infty)$. Prove that the function

$$[0, \infty) \ni t \mapsto \|X(t)^{-1}\| \in (0, \infty)$$

is also bounded.

Hint. Use Liouville's theorem.

3.2 Prove that, if all the solutions of system (3.108) are bounded on $[0, \infty)$ and (3.109) holds, then any solution of the system

$$\boldsymbol{x}' = B(t)\boldsymbol{x}, \quad t \geq 0, \tag{3.110}$$

is bounded on $[0, \infty)$. Above, $B(t)$ is an $n \times n$ matrix whose entries depend continuously on $t \geq 0$ and satisfy the condition

$$\int_0^\infty \|B(s) - A(s)\|ds < \infty.$$

Hint. Rewrite system (3.110) in the form $\boldsymbol{x}' = A(t)\boldsymbol{x} + \big(B(t) - A(t)\big)\boldsymbol{x}$, and then use the formula of variation of constants.

3.3 Prove that all the solutions of the differential equation

$$x' + \left(1 + \frac{2}{t(1+t^2)}\right)x = 0$$

are bounded on $[0, \infty)$.

Hint. Interpret the function $f(t) = -2x(t(1+t^2))^{-1}$ as a nonhomogeneous term and then use the formula of variation of constants.

3.4 Express as a power series the solution of the Cauchy problem

$$x'' - tx = 0, \quad x(0) = 0, \quad x'(0) = 1.$$

3.5 Consider the linear second-order equation

$$x'' + a_1(t)x' + a_2(t)x = 0, \quad t \in I := [\alpha, \beta], \tag{3.111}$$

where $a_i : I \to \mathbb{R}$, $i = 1, 2$, are continuous functions. A *zero* of a solution $x(t)$ is a point $t_0 \in I$ such that $x(t_0) = 0$. Prove that the following hold.

(i) The set of zeros of a nonzero solution is at most countable, and contains only isolated points.
(ii) The zero sets of two linearly independent solutions x, y separate each other, that is, between any two consecutive zeros of x there exists precisely one zero of y. (This result is due to *J. Sturm* (1803–1855).)

Hint.

(i) Follows by the uniqueness of solutions to the Cauchy problem for Eq. (3.111).
(ii) Let t_1, t_2 be two consecutive zeros of x. $y(t) \neq 0$ on $[t_1, t_2]$, and the function $\varphi(t) = \frac{x(t)}{y(t)}$ is C^1 on $[t_1, t_2]$. Use Rolle's theorem to reach a contradiction.

3.6 Prove that the equation $x'' = a(t)x$ is non-oscillatory (that is, it admits solutions with only finitely many zeros) if and only if the Riccati equation $y' = -y^2 + a(t)$ admits solutions defined on the entire semi-axis $[0, \infty)$.

Hint. Use Problem 1.14.

3.7 Consider the second-order equations

$$x'' + a(t)x = 0, \tag{3.112}$$
$$x'' + b(t)x = 0, \tag{3.113}$$

where a, b are continuous functions on an interval $I = [t_1, t_2]$. Prove that, if $\varphi(t)$ is a solution of (3.112) and $\psi(t)$ is a solution of (3.113), then we have the identity

$$\begin{vmatrix} \varphi(t_2) & \psi(t_2) \\ \varphi'(t_2) & \psi'(t_2) \end{vmatrix} - \begin{vmatrix} \varphi(t_1) & \psi(t_1) \\ \varphi'(t_1) & \psi'(t_1) \end{vmatrix} = \int_{t_1}^{t_2} \big(a(t) - b(t)\big)\varphi(t)\psi(t)dt. \tag{3.114}$$

3.8 (*Sturm's comparison theorem*) Under the same assumptions as in Problem 3.7, prove that, if $a(t) \leq b(t)$, $\forall t \in I$, then between any two consecutive zeros of the solution $\varphi(t)$ there exists at least one zero of the solution $\psi(t)$.

Hint. Use identity (3.114).

3.9 Find all the values of the complex parameter λ such that the boundary value problem

$$x'' + \lambda x = 0, \quad t \in [0, 1], \tag{3.115}$$

$$x(0) = x(1) = 0 \tag{3.116}$$

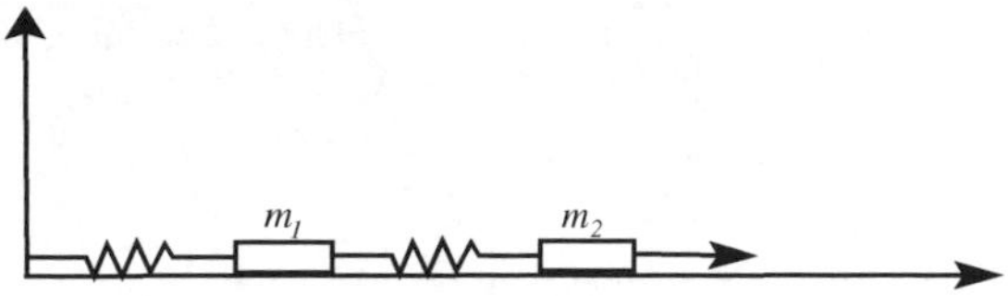

Fig. 3.3 An elastic chain

admits nontrivial solutions. (A boundary value problem as above is known as a *Sturm–Liouville problem*. The corresponding λ's are called the *eigenvalues* of the problem.)

Hint. Prove first that λ has to be a nonnegative real number. Solve (3.115) separately in the cases $\lambda < 0$, $\lambda = 0$ and $\lambda > 0$ and then impose condition (3.116) to find that $\lambda = (n\pi)^2$, $n \in \mathbb{Z}_{>0}$.

3.10 The differential system

$$\begin{aligned} m_1 x_1'' + \omega_1 x_1 - \omega_2(x_2 - x_1) &= 0, \\ m_2 x_2'' + \omega_2(x_2 - x_1) &= f \end{aligned} \tag{3.117}$$

describes the motion of a mechanical system made of two particles of masses m_1 and m_2 serially connected to a fixed point through two elastic springs with elasticity constants ω_1 and ω_2; see Fig. 3.3.

Solve the system in the special case $m_1 = m_2 = m$, $\omega_1 = \omega_2 = \omega$ and $f = 0$.

3.11 Solve the differential equation

$$\begin{aligned} x'' + \omega^2 x + \varepsilon^{-1}\min(x, 0) = 0, \quad t \geq 0, \\ x(0) = x_0, \quad x'(0) = 0, \end{aligned} \tag{3.118}$$

where $x_0 \geq 0$, $\varepsilon > 0$ and investigate the behavior of the solution as $\varepsilon \searrow 0$.

Hint. One considers separately the cases $x_0 > 0$ and $x_0 = 0$. The limit case $\varepsilon = 0$ models the harmonic motion in the presence of an obstacle at $x = 0$. In the limit case $\varepsilon = 0$, the solution formally satisfies the system

$$\begin{aligned} \big(x''(t) + \omega^2 x(t)\big)x(t) = 0, \quad \forall t \geq 0, \\ x(t) \geq 0, \quad x''(t) + \omega^2 x(t) \geq 0, \end{aligned}$$

which is a differential variational inequality of the form (2.91).

3.12 Prove that the matrix $X(t) = e^{(\ln t)A}$ is a fundamental matrix of the system $t\boldsymbol{x}' = \boldsymbol{x}$, $t > 0$.

3.13 Prove that, for any $n \times n$ matrix A and any $t \in \mathbb{R}$, we have

$$\left(e^{tA}\right)^* = e^{tA^*},$$

where * indicates the transpose of a matrix.

Hint. Use formula (3.61).

3.14 Prove that the solution $X(t)$ of the matrix-valued differential equation

$$X'(t) = AX(t) + X(t)B, \quad t \in \mathbb{R},$$

satisfying the initial condition $X(0) = C$ is given by the formula

$$X(t) = e^{tA}Ce^{tB},$$

where A, B, C are $n \times n$ matrices.

Hint. Use equation (3.8).

3.15 Prove all the entries of e^{tA} are positive for any $t \geq 0$ if and only if

$$a_{ij} \neq 0, \quad \forall i \neq j, \tag{3.119}$$

where a_{ij} are the entries of the $n \times n$ matrix A.

Hint. From formula (3.61), where $t \to 0$, we see that condition (3.119) is necessary. Conversely, if (3.119) holds, then there exists an $\alpha > 0$ such that all the entries of $\alpha\mathbb{1} + A$ are positive. Next, use the equality $e^{tA} = e^{-\alpha t}e^{t(\alpha\mathbb{1}+A)}$.

3.16 Prove that, if (3.119) holds, then the solution $\boldsymbol{x}$ of the Cauchy problem

$$\boldsymbol{x}' = A\boldsymbol{x} + \boldsymbol{f}(t), \quad \boldsymbol{x}(0) = \boldsymbol{x}_0,$$

where $\boldsymbol{x}_0 \geq 0$ and $\boldsymbol{f}(t) \geq 0$, is positive, that is, all its components are positive.

Hint. Use formula (3.21).

3.17 Prove that, if the $n \times n$ matrices A, B commute, that is, $AB = BA$, then

$$e^A e^B = e^{(A+B)} = e^B e^A.$$

Hint. Prove that $e^{(A+B)t} = e^{At}e^{Bt}$, $\forall t \geq 0$, using Theorem 3.11 and (3.62).

3.18 Let A be a constant $n \times n$ matrix.

(i) Prove that there exists a real number ω and a positive constant M such that

$$\| e^{tA} \| \leq Me^{\omega|t|}, \quad \forall t \in \mathbb{R}. \tag{3.120}$$

(ii) Prove that, for any complex numbers λ such that $\mathbf{Re}\,\lambda > \omega$, the matrix $\lambda\mathbb{1} - A$ is nonsingular and we have

$$\left(\lambda\mathbb{1} - A\right)^{-1} = \int_0^\infty e^{-\lambda t} e^{tA} dt. \tag{3.121}$$

Hint. Use formula (3.62).

3.19 Let A be an $n \times n$ matrix. Prove (Trotter's formula)

$$e^{tA} = \lim_{k\to\infty} \left(\mathbb{1} + \frac{t}{k}A\right)^k = \lim_{k\to\infty} \left(\mathbb{1} - \frac{t}{k}A\right)^{-k}, \quad \forall t \in \mathbb{R}. \tag{3.122}$$

Hint. Prove first that

$$\left(\mathbb{1} + \frac{t}{k}A\right)^k = \frac{1}{2\pi \boldsymbol{i}} \int_\Gamma \left(1 + \frac{t\lambda}{k}\right)^k \left(\lambda\mathbb{1} - A\right)^{-1} d\lambda, \tag{3.123}$$

where Γ is a contour as in the proof of Theorem 3.13. Next, let $k \to \infty$.

3.20 Let $\{A_j\}_{j\geq 1}$ be a sequence of $n \times n$ matrices that converges in norm to the $n \times n$ matrix A. Prove that

$$\lim_{j\to\infty} e^{tA_j} = e^{tA}, \quad \forall t \in \mathbb{R}. \tag{3.124}$$

Hint. Use the representation formula (3.64) taking into account that $A_j \to A$ implies that

$$(\lambda\mathbb{1} - A_j)^{-1} \to (\lambda\mathbb{1} - A)^{-1}, \ \forall\lambda \,\overline{\in}\, \operatorname{spec}(A).$$

3.21 Find the general solution of the equation

$$x^{(3)} + x = \delta_1. \tag{3.125}$$

Hint. Use the fact that $x^{(3)} + x = 0$ on $(-\infty, 1) \cup (1, +\infty)$ and proceed as in Example 3.4.

3.22 Let A be a real $n \times n$ matrix, and $D \subset \mathbb{R}^n$ be an invariant linear subspace of A, that is, $AD \subset D$. Prove that, if $\boldsymbol{x}_0 \in D$, then the solution $\boldsymbol{x}(t)$ of the Cauchy problem

$$\boldsymbol{x}' = A\boldsymbol{x}, \quad \boldsymbol{x}(0) = \boldsymbol{x}_0$$

stays in D for any $t \in \mathbb{R}$.

Hint. Use Theorem 3.11 and formula (3.62).

3.23 Let A be a real $n \times n$ matrix and B an $n \times m$ matrix. Prove that, if

$$\text{rank}\,[B, AB, A^2B, \ldots, A^{n-1}B] = n, \tag{3.126}$$

then the only vector $\boldsymbol{x} \in \mathbb{R}^n$ such that

$$B^* e^{A^* t}\boldsymbol{x} = 0, \quad \forall t \geq 0, \tag{3.127}$$

is the null vector.

Hint. Differentiating equality (3.127) and setting $t = 0$, we deduce that

$$B^*\boldsymbol{x} = B^*A^*\boldsymbol{x} = \cdots = B^*(A^*)^{n-1}\boldsymbol{x} = 0.$$

The algebraic condition (3.126) then implies that $\boldsymbol{x} = 0$. Condition (3.126) was introduced by Kalman (n. 1930) and plays an important role in the theory of controllability of linear differential systems.

3.24 Let A be an $n \times n$ matrix. Study the domain of definition of the matrix-valued function

$$\sin(tA) = \sum_{k=0}^{\infty}(-1)^k \frac{t^{2k+1}}{(2k+1)!}A^{2k+1},$$

and prove that, for any $\boldsymbol{x}_0 \in \mathbb{R}$, the function $\boldsymbol{x}(t) = \sin(tA)\boldsymbol{x}_0$ is a solution of the second-order linear differential system $\boldsymbol{x}'' + A^2\boldsymbol{x} = 0$.

Hint. Use Proposition 3.2.

3.25 Compute e^{tA} when A is one of the following matrices

$$\begin{bmatrix} 2 & -1 \\ -2 & 3 \end{bmatrix}, \quad \begin{bmatrix} -1 & 0 & 3 \\ -8 & 1 & 12 \\ -2 & 0 & 4 \end{bmatrix}, \quad \begin{bmatrix} -1 & 2 & -1 \\ -1 & -4 & 1 \\ -1 & -2 & -1 \end{bmatrix}.$$

Hint. Use (3.64).

3.26 Prove that, using the substitution $t = e^{\tau}$, the *Euler equation*

$$t^n x^{(n)} + a_1 t^{n-1} x^{(n-1)} + \cdots + a_n x = 0,$$

where $a_1, \ldots, a_n$ are real constants, reduces to a linear differential equation of order n with constant coefficients.

3.27 Prove that, if A, B are $m \times m$ real matrices, then we have *Lie's formula* (*S. Lie* (1842–1899))

$$e^{t(A+B)} = \lim_{n \to \infty}\left(e^{\frac{t}{n}A}e^{\frac{t}{n}B}\right)^n, \quad \forall t \geq 0, \ \forall t \geq 0. \tag{3.128}$$

Hint. For any positive integer n, the matrix

$$Y_n(t) = \left(e^{\frac{t}{n}A} e^{\frac{t}{n}B} \right)^n$$

satisfies the differential equation

$$Y_n'(t) = (A+B)Y_n(t) + \left(e^{\frac{t}{n}A} B e^{-\frac{t}{n}A} - B \right) Y_n(t).$$

Then, by (3.21),

$$Y_n(t) = e^{t(A+B)} + \int_0^t e^{(t-s)(A+B)} (e^{\frac{s}{n}A} B e^{-\frac{s}{n}A} - B) Y_n(s) ds,$$

from which we can conclude that $Y_n(t) \to e^{t(A+B)}$ as $n \to \infty$.

Remark 3.6 Formula (3.128) is equivalent to the convergence of the *fractional step method*

$$x_\varepsilon'(t) = Ax_\varepsilon(t),\ t \in [i\varepsilon, (i+1)\varepsilon];\ x_\varepsilon(i\varepsilon) = y_\varepsilon(\varepsilon),$$
$$y_\varepsilon'(t) = By_\varepsilon(t),\ t \in [0, \varepsilon];\ y_\varepsilon(0) = x_\varepsilon(i\varepsilon - 0),$$

to the solution x to the Cauchy problem $x' = (A+B)x$, $x(0) = x_0$.

Chapter 4
Stability Theory

The concept of stability has its origin in the problem of equilibrium of conservative mechanical systems. Generally speaking, a motion is stable if "small" perturbations of the initial conditions lead to only "small" variations of the motion over an infinite period of time.

Interest in the mathematical theory of stability of motion was stimulated by late nineteenth century research in celestial mechanics. The works of *H. Poincaré* (1854–1912) and *J.C. Maxwell* (1831–1879) represent pioneering contributions in this field. We recall that Maxwell used this concept in the study of Saturn's rings, discovering that the only configurations that are mathematically stable correspond to a discontinuous structure of the rings.

The modern theory of stability of differential systems is due to the Russian mathematician *A.M. Lyapunov* (1957–1918) and has as its starting point his doctoral dissertation, defended in 1882 and entitled "*On the general problem of the stability of motion*". In the last few decades, the theory has been enriched by many important results, some motivated by the wide range of applications of the theory of stability in the study and design of control systems.

In this chapter, we will describe only the basics of this theory, together with some illustrations on certain systems of interest in physics. Basic references for this chapter are [1, 6, 15].

4.1 The Concept of Stability

Consider the differential system

$$x' = f(t, x), \tag{4.1}$$

V. Barbu, *Differential Equations*, Springer Undergraduate Mathematics Series,
DOI 10.1007/978-3-319-45261-6_4

where $f : \Omega \to \mathbb{R}^n$ is defined on the region $\Omega = \{(t, \boldsymbol{x}) \in (0, \infty) \times \mathbb{R}^n;\ \|\boldsymbol{x}\| < a\}$. We assume that f is continuous in $(t, \boldsymbol{x})$ and locally Lipschitz in $\boldsymbol{x}$.

From the existence, uniqueness and extendibility theorems, we deduce that, for any $(t_0, \boldsymbol{x}_0) \in \Omega$, system (4.1) admits a unique right-saturated solution $\boldsymbol{x}(t; t_0, \boldsymbol{x}_0)$ defined on a maximal interval $[t_0, T)$. Suppose that $\varphi(t)$ is a solution of (4.1) defined on the entire semi-axis $[t_0, \infty)$.

Definition 4.1 The solution φ is called *stable* if, for any $\varepsilon > 0$, there exist $a > 0$ and $\delta(\varepsilon, t_0) > 0$ such that, for any $\boldsymbol{x}_0$ satisfying $\|\boldsymbol{x}_0\| < a$ and $\|\boldsymbol{x}_0 - \varphi(t_0)\| < \delta(\varepsilon, t_0)$, the following hold.

(i) The solution $\boldsymbol{x}(t; t_0, \boldsymbol{x}_0)$is defined on $[t_0, \infty)$.

(ii) $\big\| \boldsymbol{x}(t; t_0, \boldsymbol{x}_0) - \varphi(t) \big\| \le \varepsilon, \quad \forall t \ge t_0.$ (4.2)

The solution is called *uniformly stable* if the above $\delta(\varepsilon, t_0)$ can be chosen independent of t_0.

Definition 4.2 (a) The solution φ is called *asymptotically stable* if it is stable and there exists a $\mu(t_0) > 0$ such that

$$\lim_{t \to \infty} \| \boldsymbol{x}(t; t_0, \boldsymbol{x}_0) - \varphi(t) \| = 0, \tag{4.3}$$

for any $\boldsymbol{x}_0$ such that $\|\boldsymbol{x}_0 - \varphi(t_0)\| \le \mu(t_0)$.

(b) The solution φ is called *uniformly asymptotically stable* if it is uniformly stable and there exists a $\mu_0 > 0$, *independent of* t_0, such that, if $\|\boldsymbol{x}_0 - \varphi(t_0)\| \le \mu_0$, then

$$\lim_{t \to \infty} \| \boldsymbol{x}(t; t_0, \boldsymbol{x}_0) - \varphi(t) \| = 0,$$

uniformly with respect to t_0.

Roughly speaking, stability means continuity and low sensitivity of solutions on an infinite interval of time with respect to the initial data.

In many concrete situations, one is interested in *stationary* (or equilibrium) solutions of the differential systems, that is, solutions constant in time of the form $\varphi(t) \equiv \boldsymbol{c}$. These solutions describe the stationary regimes of various mechanisms and processes. Since, in practical situations, the equilibrium can only be determined approximatively, the only physically significant equilibrium states are the stable ones.

Using a simple algebraic trick, we can reduce the study of the stability of a solution $\varphi(t)$ of system (4.1) to the study of the stability of the trivial (or null) solution $\boldsymbol{x} \equiv 0$ of a different system. Indeed, the substitution $\boldsymbol{y} := \boldsymbol{x} - \varphi$ in (4.1) leads us to the differential system

$$\boldsymbol{y}' = \boldsymbol{f}(t, \boldsymbol{y} + \varphi) - \varphi', \quad t \ge 0. \tag{4.4}$$

The solution φ of (4.1) corresponds to the trivial solution of system (4.4). Thus, without loss of generality, we can restrict our attention to the stability of the trivial solution. For this reason, in the sequel, we will assume that the function f satisfies the additional condition

$$f(t,0)=0, \quad \forall t \geq t_0, \tag{4.5}$$

which amounts to saying that $\boldsymbol{x} \equiv 0$ is a solution of (4.1). In this case, Definition 4.1 for $\varphi = 0$ takes the following form.

Definition 4.3 The trivial solution $\varphi = 0$ is called *stable* if for any $\varepsilon > 0$ there exist $a > 0$ and $\delta(\varepsilon, t_0) > 0$ such that, for any $\boldsymbol{x}_0$ satisfying $\|\boldsymbol{x}_0\| < a$ and $\|\boldsymbol{x}_0 - \varphi(t_0)\| < \delta(\varepsilon, t_0)$, the following hold.

(i) The solution $\boldsymbol{x}(t; t_0, \boldsymbol{x}_0)$ is defined on $[t_0, \infty)$.

(ii) $\big\| \boldsymbol{x}(t; t_0, \boldsymbol{x}_0) \big\| \leq \varepsilon, \quad \forall t \geq t_0.$ (4.6)

The trivial solution is called *asymptotically stable* if it is stable and there exists a $\mu(t_0) > 0$ such that

$$\lim_{t\to\infty} \| \boldsymbol{x}(t; t_0, \boldsymbol{x}_0) \| = 0, \tag{4.7}$$

for any $\boldsymbol{x}_0$ such that $\|\boldsymbol{x}_0\| \leq \mu(t_0)$.

Given the n-th order differential equation

$$x^{(n)} = g\big(t, x, x', \dots, x^{(n-1)}\big), \tag{4.8}$$

we will say that its trivial solution (if it exists) is *stable* if the trivial solution of the associated system is stable

$$\begin{aligned} x_1' &= x_2 \\ x_2' &= x_3 \\ \vdots \;\; & \;\vdots \;\; \vdots \\ x_n' &= g(t, x_1, \dots, x_n). \end{aligned} \tag{4.9}$$

Remark 4.1 We want to emphasize that stability is a property of a *solution*, not of the system. It is possible that the same system has both stable and unstable solutions. Take, for example, the pendulum equation

$$x'' + \sin x = 0, \quad t \geq 0. \tag{4.10}$$

This equation admits two stationary solutions, $\varphi_1(t) = 0$, and $\varphi_2(t) = \pi$, corresponding to the two equilibrium positions of the pendulum.

The solution φ_1 is stable. Indeed, multiplying (4.10) by x' and integrating, we find that

$$|x'(t)|^2 - 2\cos x(t) = |x'(0)|^2 - 2\cos x(0), \quad \forall t \geq 0, \tag{4.11}$$

for any solution $x(t)$ of (4.10). If $|x'(0)| + |x(0)| < \delta$, then (4.11) implies that

$$|x'(0)|^2 \leq 2\big(\cos\delta - 1\big) + \delta^2 + 2\big(1 - \cos\delta\big), \quad \forall t \geq 0.$$

Given $\varepsilon > 0$, we choose $\delta = \delta(\varepsilon) > 0$ such that

$$\delta^2 + 2(1 - \cos\delta) \leq \varepsilon^2,$$

and we deduce that

$$|x'(t)|^2 + 4\sin\frac{x^2(t)}{2} \leq \varepsilon^2, \quad \forall t \geq 0.$$

This shows that the trivial solution to Eq. (4.10) is stable. (We ought to point out that the trivial solution is not asymptotically stable.)

To study the stability of the solution φ_2, we use the substitution $y := x - \pi$ and reduce the problem to the study of the trivial solution of

$$y'' - \sin y = 0. \tag{4.12}$$

Let $y(t)$ be an arbitrary solution of the above equation. Arguing as above, we obtain the equality

$$|y'(t)|^2 + 2\cos y(t) = |y'(0)|^2 + 2\cos y(0), \quad \forall t \geq 0. \tag{4.13}$$

Consider now the solution of the Cauchy problem with the initial conditions

$$y'(0) = \delta, \quad y(0) = 2\arcsin\frac{\delta}{2}.$$

Then

$$\cos y(0) = 1 - 2\sin^2\arcsin\frac{\delta}{2} = 1 - \frac{\delta^2}{2}, \quad |y'(0)|^2 + 2\cos y(0) = 2.$$

Using (4.13), we deduce that

$$\big(y'(t)\big)^2 = 2\big(1 - \cos y(t)\big)^2 = 4\sin^2\frac{y(t)}{2}.$$

This implies that $y(t)$ is the solution of the Cauchy problem

$$y' = 2\sin\frac{y}{2}, \quad y(0) = 2\arcsin\frac{\delta}{2}.$$

This equation can be integrated and we deduce that

$$y(t) = 4\arctan(Ce^t), \quad C = \tan\frac{y(0)}{4}.$$

Thus, for $\varepsilon > 0$, there exists no $\delta > 0$ such that

$$|y(t)|,\ |y'(t)| \leq 1,\quad \forall t \geq 0,$$

and

$$|y(0)|,\quad |y'(0)| \leq \delta.$$

These conclusions are in perfect agreement with observed reality.

Definition 4.4 If any solution of system (4.1) is defined on $[0, \infty)$ and converges to 0 as $t \to \infty$, then we say that system (4.1) is *globally asymptotically stable.*

In the next section, we will encounter many examples of globally asymptotically stable systems.

4.2 Stability of Linear Differential Systems

Consider the linear homogeneous system

$$\boldsymbol{x}' = A(t)\boldsymbol{x},\ t \geq 0, \tag{4.14}$$

where $A(t)$ is an $n \times n$ matrix whose entries are continuous functions $[0, \infty) \to \mathbb{R}$.

Proposition 4.1 *If the trivial solution of system* (4.14) *is stable* (*respectively uniformly stable, or asymptotically stable*), *then any other solution of this system is stable* (*respectively uniformly stable or asymptotically stable*).

Proof If $\boldsymbol{x} = \varphi(t)$ is an arbitrary solution of the system, then via the substitution $y := \boldsymbol{x} - \varphi(t)$ we reduce it to the trivial solution $\boldsymbol{y} = 0$ of the same linear system. The stability of $\varphi(t)$ is thus identical to the stability of the trivial solution. □

We deduce that, *in the linear case*, the stability property of the trivial solution is a *property of the system.* In other words, all solutions of this system enjoy the same stability properties: a solution is stable if and only if all the solutions are stable, etc. We will say that system (4.14) is *stable* (respectively *asymptotically stable*) if all the solutions of this system are such. The central result of this section is the following characterization of the stability of linear differential systems.

Theorem 4.1 (a) *The linear differential system* (4.14) *is stable if and only if there exists a fundamental matrix $X(t)$ of the system which is bounded on* $[0, \infty)$. *If this happens, all the fundamental matrices are bounded on* $[0, \infty)$.
(b) *The linear differential system* (4.14) *is asymptotically stable if and only if there exists a fundamental matrix $X(t)$ such that*

$$\lim_{t \to \infty} \|X(t)\| = 0.$$

If such a fundamental matrix exists, then all the fundamental matrices satisfy the above property. Above, $\|X\|$ denotes the norm of the matrix X defined as in Appendix A.

Proof (a) Suppose first that system (4.14) admits a fundamental solution $X(t)$ that is bounded on the semi-axis $[0, \infty)$, that is,

$$\exists M > 0 : \quad \|X(t)\| \leq M, \quad \forall t \geq 0.$$

For any $\boldsymbol{x}_0 \in \mathbb{R}$, and any $t_0 \geq 0$, the solution $\boldsymbol{x}(t; t_0, \boldsymbol{x}_0)$ is given by (3.9)

$$\boldsymbol{x}(t; t_0, \boldsymbol{x}_0) = X(t)X(t_0)^{-1}\boldsymbol{x}_0.$$

Hence

$$\|\boldsymbol{x}(t; t_0, \boldsymbol{x}_0)\| \leq \|X(t)\| \cdot \|X(t_0)^{-1}\| \cdot \|\boldsymbol{x}_0\| \leq M\|X(t_0)^{-1}\| \cdot \|\boldsymbol{x}_0\|, \tag{4.15}$$

$\forall t \geq 0$. Thus

$$\| \boldsymbol{x}(t; t_0, \boldsymbol{x}_0) \| \leq \varepsilon, \quad \forall t \geq 0,$$

as soon as

$$\|\boldsymbol{x}_0\| \leq \delta(\varepsilon) := \frac{\varepsilon}{M\|X(t_0)^{-1}\|}.$$

Conversely, let us assume that the trivial solution 0 is stable. Let $X(t)$ denote the fundamental matrix determined by the initial condition $X(0) = \mathbb{1}$. Since the trivial solution is stable, we deduce that there exists a $\delta > 0$ such that, for any $\boldsymbol{x}_0 \in \mathbb{R}$ satisfying $\|\boldsymbol{x}_0\| \leq \delta$, we have

$$\|X(t)\boldsymbol{x}_0\| = \|\boldsymbol{x}(t; 0, \boldsymbol{x}_0)\| \leq 1, \quad \forall t \geq 0.$$

We deduce that

$$\|X(t)\| \leq \frac{1}{\delta}, \quad \forall t \geq 0.$$

Since any other fundamental matrix $Y(t)$ is related to $X(t)$ by a linear equation

$$Y(t) = X(t)C, \quad \forall t \geq 0,$$

where C is a nonsingular, time-independent $n \times n$ matrix, we deduce that $Y(t)$ is also bounded on $[0, \infty)$.

Part (b) is proved in a similar fashion using estimate (4.15), where the constant M is replaced by a positive function $M(t)$, such that $\lim_{t\to\infty} M(t) = 0$. □

Consider now the case when $A(t) \equiv A$ is independent of t. A matrix A is called *Hurwitzian* if all its eigenvalues have negative real parts. Theorem 4.1 implies the following criterion of asymptotic stability for linear systems with constant coefficients.

Theorem 4.2 *Let A be a real $n \times n$ matrix. Then the linear system*

$$\boldsymbol{x}' = A\boldsymbol{x} \tag{4.16}$$

is asymptotically stable if and only if A is a Hurwitzian matrix.

Proof According to Theorem 4.1, system (4.16) is asymptotically stable if and only if

$$\lim_{t\to\infty} \|e^{tA}\| = 0.$$

On the other hand, Theorem 3.12 shows that, for any $1 \leq i, j \leq n$, the (i, j)-entry of e^{tA} has the form

$$\sum_{\lambda\in\mathrm{spec}(A)} p_{i,j,\lambda}(t)e^{t\lambda}, \tag{4.17}$$

where, for any eigenvalue $\lambda \in \mathrm{spec}(A)$, we denote by $p_{i,j,\lambda}(t)$ a polynomial in t of degree smaller than the algebraic multiplicity of λ. The above equality shows that

$$\lim_{t\to\infty} \|e^{tA}\| = 0 \quad \text{if and only if} \quad \mathbf{Re}\,\lambda < 0, \quad \forall \lambda \in \mathrm{spec}(A). \qquad \square$$

Corollary 4.1 *If A is a Hurwitzian matrix, then system* (4.16) *is asymptotically stable. Moreover, for any positive number ω such that*

$$\omega < \min\big\{ -\mathbf{Re}\,\lambda;\quad \lambda \in \mathrm{spec}(A)\big\},$$

there exists an $M > 0$ such that, for any $\boldsymbol{x}_0 \in \mathbb{R}$ and any $t_0 \geq 0$, we have

$$\|\boldsymbol{x}(t; t_0, \boldsymbol{x}_0)\| \leq Me^{-\omega(t-t_0)}, \quad \forall t \geq t_0. \tag{4.18}$$

Proof The asymptotic stability statement follows from Theorem 4.2. Next, observe that

$$\boldsymbol{x}(t; t_0, \boldsymbol{x}_0) = e^{(t-t_0)A}\boldsymbol{x}_0.$$

Estimate (4.18) now follows from the structural equalities (4.17). $\square$

From the structural equalities (4.17), we obtain the following stability criterion.

Corollary 4.2 *If all the eigenvalues of A have nonpositive real parts, and the ones with trivial real parts are simple, then system* (4.16) *is stable.*

To be able to apply Theorem 4.2 in concrete situations and for differential systems of dimensions ≥ 3, we need to know criteria deciding when a polynomial is

Hurwitzian, that is, all its roots have negative real parts. Such a criterion was found by *A. Hurwitz* (1859–1919) and can be found in many classical algebra books, e.g. [10, Chap. XV, Sect. 6]. For degree 3 polynomials

$$p(\lambda) = \lambda^3 + a_1\lambda^2 + a_2\lambda + a_3$$

it reads as follows: *the polynomial $p(\lambda)$ is Hurwitzian if and only if*

$$a_1 > 0, \quad a_3 > 0, \quad a_1a_2 > a_3.$$

Finally, let us comment on higher order linear differential equations with constant coefficients

$$x^{(n)} + a_1x^{(n-1)} + \cdots + a_nx = 0. \tag{4.19}$$

Taking into account the equivalence between such equations and linear differential systems of dimension n with constant coefficients and the fact that the eigenvalues of the associated matrix are the roots of the characteristic equation

$$\lambda^n + a_1\lambda^{n-1} + \cdots + a_n = 0, \tag{4.20}$$

we obtain from Theorem 4.2 the following result.

Corollary 4.3 *The trivial solution of* (4.19) *is asymptotically stable if and only if all the roots of the characteristic equation* (4.20) *have negative real parts.*

The previous discussion shows that, if the trivial solution of Eq. (4.19) is asymptotically stable, then all the other solutions converge exponentially to 0 as $t \to \infty$, together with all their derivatives of orders $\leq n$.

4.3 Stability of Perturbed Linear Systems

In this section, we will investigate the stability of the trivial solutions of the differential systems of the form

$$\boldsymbol{x}' = A\boldsymbol{x} + \boldsymbol{F}(t, \boldsymbol{x}), \quad t \geq 0, \tag{4.21}$$

where A is a fixed $n \times n$ real matrix and $\boldsymbol{F} : [0, \infty) \times \mathbb{R}^n \to \mathbb{R}^n$ is a function continuous in the cylinder

$$\Omega = \big\{(t, \boldsymbol{x}) \in [0, \infty) \times \mathbb{R}^n;\ \ \|\boldsymbol{x}\| < r\big\},$$

locally Lipschitz in $\boldsymbol{x}$, and such that $\boldsymbol{F}(t, 0) \equiv 0$. Such a system is called a *perturbed linear system* and $\boldsymbol{F}$ is called a *perturbation*.

Systems of the form (4.21) arise naturally when linearizing arbitrary differential systems at the trivial systems. For sufficiently small perturbations $\boldsymbol{F}$, we expect the stability, or asymptotic stability, of the trivial solution of the linear system to "propagate" to the perturbed systems as well. This is indeed the case, and the next results states this in a precise fashion.

Theorem 4.3 (Lyapunov–Poincaré) *Suppose that A is a Hurwitzian matrix. Fix $M, \omega > 0$ such that*

$$\|e^{tA}\| \leq Me^{-\omega t}, \quad \forall t \geq 0. \tag{4.22}$$

If there exists an $L > 0$ such that

$$L < \frac{\omega}{M} \tag{4.23}$$

and

$$\| \boldsymbol{F}(t, \boldsymbol{x}) \| \leq L\|\boldsymbol{x}\|, \quad \forall (t, \boldsymbol{x}) \in \Omega, \tag{4.24}$$

then the trivial solution of system (4.21) *is asymptotically stable.*

Proof Fix $(t_0, \boldsymbol{x}_0) \in \Omega$ and denote by $\boldsymbol{x}(t; t_0, \boldsymbol{x}_0)$ the right saturated solution $\boldsymbol{x}(t)$ of (4.21) satisfying the initial condition $\boldsymbol{x}(t_0) = \boldsymbol{x}_0$. Let $[t_0, T)$ denote the right-maximal existence interval of this solution. Interpreting (4.21) as a nonhomogeneous linear system and using the formula of variation of constants (3.21), we deduce that

$$\boldsymbol{x}(t; t_0, \boldsymbol{x}_0) = e^{(t-t_0)A}\boldsymbol{x}_0 + \int_{t_0}^{t} e^{(t-s)A}\boldsymbol{F}\big(s, \boldsymbol{x}(s; t_0, \boldsymbol{x}_0)\big)ds, \quad \forall t \in [t_0, T). \tag{4.25}$$

Taking the norm of both sides, and using (4.22) and (4.24), we deduce that that

$$\|\boldsymbol{x}(t; t_0, \boldsymbol{x}_0)\| \leq Me^{-(t-t_0)\omega\|}\|\boldsymbol{x}_0\| + ML\int_{t_0}^{t} e^{-(t-s)\omega}\|\boldsymbol{x}(s; t_0, \boldsymbol{x}_0)\|ds, \quad \forall t \in [t_0, T). \tag{4.26}$$

We set

$$y(t) := e^{t\omega}\|\boldsymbol{x}(t; t_0, \boldsymbol{x}_0)\|.$$

From inequality (4.26), we get

$$y(t) \leq Me^{t_0\omega}\|\boldsymbol{x}_0\| + LM\int_{t_0}^{t} y(s)ds, \quad \forall t \in [t_0, T).$$

Invoking Gronwall's Lemma, we deduce that

$$y(t) \leq M\|\boldsymbol{x}_0\|e^{LM(t-t_0)+\omega t_0},$$

so that

$$\|\boldsymbol{x}(t; t_0, \boldsymbol{x}_0)\| \leq M\|\boldsymbol{x}_0\|e^{(LM-\omega)(t-t_0)}, \quad \forall t \in [t_0, T). \tag{4.27}$$

We set $\delta = \omega - LM$ and we observe that (4.23) implies $\delta > 0$. We deduce that

$$\|\boldsymbol{x}(t; t_0, \boldsymbol{x}_0)\| \leq Me^{-\delta(t-t_0)}\|\boldsymbol{x}_0\|, \quad \forall t \in [t_0, T). \tag{4.28}$$

From the above estimate, we deduce that, if $\|\boldsymbol{x}_0\| < \frac{r}{2M}$,

$$\|\boldsymbol{x}(t; t_0, \boldsymbol{x}_0)\| < \frac{r}{2}, \quad \forall t \in [t_0, T).$$

Thus the solution $\boldsymbol{x}(t; t_0, \boldsymbol{x}_0)$ stays in the interior of the cylinder Ω on its right-maximal existence interval. Invoking Theorem 2.10, we deduce that $T = \infty$. The asymptotic stability of the trivial solution now follows immediately from (4.28). □

Theorem 4.4 *Suppose that A is a Hurwitzian matrix and the perturbation $\boldsymbol{F}$ satisfies the condition*

$$\|\boldsymbol{F}(t, \boldsymbol{x})\| \leq L\big(\|\boldsymbol{x}\|\big) \cdot \|\boldsymbol{x}\|, \tag{4.29}$$

where $L(r) \to 0$ as $r \searrow 0$. Then the trivial solution of system (4.21) *is asymptotically stable.*

Proof Choose $r_0 > 0$ such that $L(r) < \frac{M}{\omega}$, $\forall 0 < r < r_0$, where M, ω are as in (4.22). The desired conclusion follows by applying Theorem 4.3 to the restriction of $\boldsymbol{F}$ to the cylinder $\Omega =]0, \infty[\times \big\{ \|\boldsymbol{x}\| < r_0 \big\}$. □

Theorem 4.4 is the basis of the so-called *first-order approximation* method of investigating the stability of the solutions of differential systems.

Consider the autonomous differential system

$$\boldsymbol{x}' = \boldsymbol{f}(\boldsymbol{x}), \tag{4.30}$$

where

$$\boldsymbol{f} : D \to \mathbb{R}^n, \quad D := \big\{\boldsymbol{x} \in \mathbb{R}^n; \ \ \|\boldsymbol{x}\| < a\big\},$$

is a C^1-map. We assume additionally that $\boldsymbol{f}(0) = 0$ and

$$\text{the Jacobian matrix } \boldsymbol{f}_{\boldsymbol{x}}(0) \text{ is Hurwitzian.} \tag{4.31}$$

Theorem 4.5 *Under the above assumptions, the trivial solution of* (4.30) *is asymptotically stable.*

Proof Since $\boldsymbol{f}$ is C^1, we have the equality

$$\boldsymbol{f}(x) = \boldsymbol{f}(0) + \boldsymbol{f}_{\boldsymbol{x}}(0)\boldsymbol{x} + \boldsymbol{F}(\boldsymbol{x}) =: A\boldsymbol{x} + \boldsymbol{F}(\boldsymbol{x}), \tag{4.32}$$

where $\|\boldsymbol{F}(\boldsymbol{x})\| \leq L\big(\|\boldsymbol{x}\| \big) \cdot \|\boldsymbol{x}\|$, $L(r) := \sup_{\|\boldsymbol{\theta}\| \leq r} \|\boldsymbol{f}_{\boldsymbol{x}}(\boldsymbol{\theta}) - \boldsymbol{f}_{\boldsymbol{x}}(0)\|$.

The stated conclusion is now a direct consequence of Theorem 4.4. □

Remark 4.2 Theorem 4.5 admits an obvious generalization to non-autonomous systems.

Example 4.1 We illustrate the general results on a second-order ODE (see L. Pontryagin, [17]),

$$Lx'' + Rx' + C^{-1}x = C^{-1}f(x'), \tag{4.33}$$

which describes the behavior of an oscillatory electrical circuit with resistance R, capacitance C and inductance L that has a nonlinear perturbation $f(x')$, (explicitly, a triode).

Equation (4.33) is equivalent to the system

$$\begin{aligned} x' &= y \\ y' &= (CL)^{-1}f(y) - (CL)^{-1}x - RL^{-1}y. \end{aligned} \tag{4.34}$$

The above system admits the stationary solution

$$x = f(0), \quad y = 0. \tag{4.35}$$

We are interested in the stability of this solution. Making the change of variables

$$x_1 := x - f(0), \quad x_2 := y,$$

we are led to the study of the trivial solution of the system

$$\begin{aligned} x_1' &= x_2 \\ x_2' &= (CL)^{-1}\big(f(x_2) - f(0) - x_1\big) - RL^{-1}x_2. \end{aligned} \tag{4.36}$$

The linearization at 0 of this system is described by the matrix

$$A = \frac{1}{CL}\begin{bmatrix} 0 & CL \\ -1 & f'(0) - RC \end{bmatrix}.$$

This matrix is Hurwitzian if and only if

$$f'(0) < RC. \tag{4.37}$$

According to Theorem 4.5, condition (4.37) implies the asymptotic stability of the stationary solution $x = f(0)$ of Eq. (4.33).

4.4 The Lyapunov Function Technique

In this section, we will describe a general method for investigating stability, known in the literature as the *Lyapunov function technique*. It can be used without having to solve the corresponding system of ODEs.

The real function V, defined on the cylinder

$$\Omega = \left\{ (t, \boldsymbol{x}) \in \mathbb{R} \times \mathbb{R}^n; \quad t > 0, \quad \|\boldsymbol{x}\| < a \right\},$$

is called *positive definite* if there exists a continuous, nondecreasing function

$$\omega : [0, \infty) \to [0, \infty) \tag{4.38}$$

such that

$$\omega(0) = 0, \quad \omega(r) > 0, \quad \forall r > 0,$$

and

$$V(t, \boldsymbol{x}) \geq \omega(\|\boldsymbol{x}\|), \quad \forall (t, \boldsymbol{x}) \in \Omega. \tag{4.39}$$

We say that V is *negative definite* if $-V$ is positive definite.

Remark 4.3 The function ω in the definition of positive definiteness satisfies the following property:

$$\begin{gathered}\forall \varepsilon > 0, \exists \delta = \delta(\varepsilon) > 0 \text{ such that, for any } r > 0 \text{ satisfying } \omega(r) < \delta, \\ \text{we have } r < \varepsilon.\end{gathered} \tag{4.40}$$

Definition 4.5 The function $V : \Omega \to [0, \infty)$ is called a *Lyapunov function* of system (4.1) if it satisfies the following conditions.

(i) The function V is C^1 on Ω.
(ii) The function V is positive definite and $V(t, 0) = 0, \forall t > 0$.
(iii) $\partial_t V(t, \boldsymbol{x}) + \big(\text{grad}_{\boldsymbol{x}} V(t, \boldsymbol{x}), \boldsymbol{f}(t, \boldsymbol{x})\big) \leq 0, \quad \forall (t, \boldsymbol{x}) \in \Omega.$

Above, $(-, -)$ denotes the canonical Euclidean scalar product on $\mathbb{R}^n$ and $\text{grad}_{\boldsymbol{x}} V$ is the gradient of the function of $\boldsymbol{x}$, that is,

$$\text{grad}_x V := \begin{bmatrix} \partial_{x_1} V \\ \vdots \\ \partial_{x_n} V \end{bmatrix}.$$

The main result of this section, known as *Lyapunov's stability theorem*, is the following.

Theorem 4.6 (Lyapunov stability) *Consider system* (4.1) *satisfying the general conditions in Sect.* 4.1.

(a) *If system* (4.1) *admits a Lyapunov function* $V(t, \boldsymbol{x})$*, then the trivial solution of this system is stable.*

(b) *Suppose that system* (4.1) *admits a Lyapunov function* $V(t, \boldsymbol{x})$ *such that the function*

$$W(t, \boldsymbol{x}) := \partial_t V(t, \boldsymbol{x}) + \big(\operatorname{grad}_x V(t, \boldsymbol{x}), \boldsymbol{f}(t, \boldsymbol{x})\big)$$

is negative definite on Ω *and*

$$V(t, \boldsymbol{x}) \leq \mu(\|\boldsymbol{x}\|), \quad \forall (t, \boldsymbol{x}) \in \Omega, \tag{4.41}$$

where $\mu : [0, \infty) \to [0, \infty)$ *is a continuous function which vanishes at the origin. Then the trivial solution of* (4.1) *is asymptotically stable.*

Proof (a) For $(t_0, \boldsymbol{x}_0) \in \Omega$, consider the right-saturated solution $\boldsymbol{x}(t; t_0, \boldsymbol{x}_0)$ of (4.1) satisfying the initial condition $\boldsymbol{x}(t_0) = \boldsymbol{x}_0$. Let $[t_0, T)$ denote the right-maximal existence interval of this solution. Using property (iii) of a Lyapunov function, we deduce the inequality

$$\frac{d}{dt} V\big(t, \boldsymbol{x}(t; t_0, \boldsymbol{x}_0)\big) \leq 0, \quad \forall t \in [t_0, T).$$

Integrating, we deduce that

$$\omega\big(\|\boldsymbol{x}(t; t_0, \boldsymbol{x}_0)\|\big) \leq V\big(t, \boldsymbol{x}(t; t_0, \boldsymbol{x}_0)\big) \leq V(t_0, \boldsymbol{x}_0), \quad \forall t \in [t_0, T). \tag{4.42}$$

Using Remark 4.3 and the continuity of V, we deduce that for any $\varepsilon > 0$ there exists an $r = r(\varepsilon) > 0$ such that, for any $\boldsymbol{x}_0$ satisfying $\|\boldsymbol{x}_0\| < r(\varepsilon)$, we have

$$\omega\big(\|\boldsymbol{x}(t; t_0, \boldsymbol{x}_0)\|\big) \leq V(t_0, \boldsymbol{x}_0) < \delta(\varepsilon), \quad \forall t \in [t_0, T),$$

where $\delta(\varepsilon)$ is defined in (4.40). Thus

$$\|\boldsymbol{x}(t; t_0, \boldsymbol{x}_0)\| < \varepsilon, \quad \forall t \in [t_0, T), \quad \forall \|\boldsymbol{x}_0\| < r(\varepsilon). \tag{4.43}$$

If we choose $\varepsilon < \frac{a}{2}$, we deduce that, if the initial condition $\boldsymbol{x}_0$ satisfies $\|\boldsymbol{x}_0\| < r(\varepsilon)$, then the solution $\boldsymbol{x}(t; t_0, \boldsymbol{x}_0)$ stays inside Ω throughout its existence interval. This shows that $T = \infty$. Since, in (4.43), the parameter $\varepsilon > 0$ can be chosen arbitrarily, we deduce that the trivial solution of (4.1) is stable.

(b) Assume now that there exists an increasing function λ that is continuous and vanishes at the origin, such that

$$\partial_t V(t, \boldsymbol{x}) + \big(\operatorname{grad}_x V(t, \boldsymbol{x}), \boldsymbol{f}(t, \boldsymbol{x})\big) \leq -\lambda(\|\boldsymbol{x}\|), \quad \forall (t, \boldsymbol{x}) \in \Omega, \tag{4.44}$$

and, moreover, inequality (4.41) holds. We deduce that

$$\frac{d}{dt}V\big(t, \boldsymbol{x}(t; t_0, \boldsymbol{x}_0)\big) + \lambda\big(\|\boldsymbol{x}(t; t_0, \boldsymbol{x}_0)\| \big) \le 0, \quad \forall t \ge t_0. \tag{4.45}$$

Integrating, we deduce that

$$V\big(t, \boldsymbol{x}(t; t_0, \boldsymbol{x}_0)\big) + \int_{t_0}^{t} \lambda\big(\|\boldsymbol{x}(s; t_0, \boldsymbol{x}_0)\| \big)ds \le V(t_0, \boldsymbol{x}_0), \quad \forall t \ge t_0. \tag{4.46}$$

From (4.45) and (4.46), it follows that the limit

$$\ell = \lim_{t\to 0} V\big(t, \boldsymbol{x}(t; t_0, \boldsymbol{x}_0)\big)$$

exists, is finite, and the function $t \mapsto \lambda\big(\|\boldsymbol{x}(t; t_0, \boldsymbol{x}_0)\|\big)$ is integrable on the semi-axis $[t_0, \infty)$. Hence, there exists a sequence $t_n \to \infty$ such that

$$\lim_{n\to\infty} \lambda\big(\|\boldsymbol{x}(t_n; t_0, \boldsymbol{x}_0)\|\big) = 0.$$

In other words, $\|\boldsymbol{x}(t_n; t_0, \boldsymbol{x}_0)\| \to 0$. From (4.41), we get

$$\lim_{n\to\infty} V\big(\boldsymbol{x}(t_n; t_0, \boldsymbol{x}_0)\big) = 0,$$

so that $\ell = 0$. Using (4.39) and (4.40), we deduce that

$$\lim_{t\to\infty} \|\boldsymbol{x}(t; t_0, \boldsymbol{x}_0)\| = 0.$$

This completes the proof of Theorem 4.6. □

Theorem 4.7 *If* $\Omega = (0, \infty) \times \mathbb{R}^n$, *and the function* ω *in* (4.39) *has the property*

$$\lim_{r\to\infty} \omega(r) = \infty, \tag{4.47}$$

then, under the same assumptions as in Theorem 4.6 (b), *the trivial solution of system* (4.1) *is globally asymptotically stable.*

Proof Let $\boldsymbol{x}_0 \in \mathbb{R}^n$ and set

$$C := \sup\big\{r \ge 0;\ \ \omega(r) \le V(t_0, \boldsymbol{x}_0)\big\}.$$

From (4.47), we see that $C < \infty$, while (4.42) implies that

$$\|\boldsymbol{x}(t; t_0, \boldsymbol{x}_0)\| < \varepsilon, \quad \forall t \in [t_0, T).$$

This proves that $T = \infty$. Using inequality (4.46), we obtain as in the proof of Theorem 4.6 that

$$\lim_{t\to\infty} \|\boldsymbol{x}(t; t_0, \boldsymbol{x}_0)\| = 0.$$

For autonomous systems

$$x' = f(x), \tag{4.48}$$

we can look for time-independent Lyapunov functions. More precisely, if

$$f : D = \big\{ x \in \mathbb{R}^n;\ \ \|x\| < a \le \infty \big\} \to \mathbb{R}^n$$

is locally Lipschitz, we define, in agreement with Definition 4.4, a Lyapunov function on D to be a function $V : D \to \mathbb{R}$ satisfying the following conditions.

(L_1) $V \in C^1(D)$, $V(0) = 0$.

(L_2) $V(x) > 0$, $\forall x \neq 0$.

(L_3) $\big(\operatorname{grad} V(x), f(x) \big) \le 0$, $\forall x \in D$.

Let us observe that condition (L_2) is equivalent to the fact that V is positive definite on any domain of the form

$$D_0 := \big\{ x \in \mathbb{R}^n;\ \ \|x\| \le b < a \big\}.$$

Indeed, if V satisfies (L_2), then

$$V(x) \ge \omega(\|x\|),\ \ \forall x \in D_0,$$

where $\omega : [0, \infty) \to [0, \infty)$ is defined by

$$\omega(r) := \begin{cases} \displaystyle\inf_{r \le \|x\| \le b} V(x), & 0 \le r \le b, \\ \omega(b), & r > b. \end{cases}$$

One can easily check that $\omega(r)$ is continuous, nondecreasing, and satisfies (4.38) and (4.39). We should also mention that, in this special case, assumption (4.41) is automatically satisfied with $\mu : [0, \infty) \to [0, \infty)$ given by

$$\mu(r) := \begin{cases} \displaystyle\sup_{\|x\| \le r} V(x), & 0 \le r \le b, \\ \mu(b), & r > b. \end{cases}$$

□

Theorems 4.6 and 4.7 have the following immediate consequence.

Corollary 4.4 *If there exists a function V satisfying (L_1), (L_2), (L_3), then the trivial solution of* (4.48) *is stable. If, additionally, V satisfies*

$$\big(\operatorname{grad} V(x), f(x) \big) < 0,\ \ \forall x \in D \setminus \{0\}, \tag{4.49}$$

then the trivial solution is asymptotically stable. Furthermore, if V is coercive, *that is,*

$$\lim_{\|\boldsymbol{x}\| \to \infty} V(\boldsymbol{x}) = \infty,$$

then the trivial solution is globally asymptotically stable.

Remark 4.4 It is worth remembering that, for time-independent functions V, the positivity condition (4.39) is equivalent to the condition $V(\boldsymbol{x}) > 0$, $\forall \boldsymbol{x} \neq 0$. This is easier to verify in concrete situations.

Example 4.2 The Lyapunov function technique can sometime clarify situations that are undecidable using the first approximation method. Consider, for example, the differential system

$$\begin{aligned} x_1' &= 2x_2(x_3 - 2), \\ x_2' &= -x_1(x_3 - 1), \\ x_3' &= x_1 x_2. \end{aligned}$$

We want to investigate the stability of the trivial solution (0, 0, 0). The linearization at the origin of this system is described by the matrix

$$\begin{bmatrix} 0 & -4 & 0 \\ 1 & 0 & 0 \\ 0 & 0 & 0 \end{bmatrix}$$

whose eigenvalues are

$$\lambda_1 = 0, \quad \lambda_2 = 2\boldsymbol{i}, \quad \lambda_3 = -2\boldsymbol{i}.$$

Hence, this matrix is not Hurwitzian and thus we cannot apply Theorem 4.5.

We seek a Lyapunov function of the form

$$V(x_1, x_2, x_3) = \alpha x_1^2 + \beta x_2^2 + \gamma x_3^2,$$

where α, β, γ are positive constants. Then

$$\operatorname{grad} V = (2\alpha x_1, \ 2\beta x_2, \ 2\gamma x_3).$$

If we denote by $\boldsymbol{f}$ the vector defining the right-hand side of our system, we deduce that

$$\big(\operatorname{grad} V, \boldsymbol{f}\big) = 4\alpha x_1 x_2 (x_3 - 2) - 2\beta x_1 x_2 (x_3 - 1) + 2\gamma x_1 x_2 x_3.$$

Note that, if

$$\alpha = \frac{1}{2}, \quad \beta = 2, \quad \gamma = 1,$$

then

$$\big(\operatorname{grad} V, \boldsymbol{f}\big) = 0 \text{ in } \mathbb{R}^3,$$

and thus V is a Lyapunov function of the system and, therefore, the trivial solution is stable.

Theorem 4.8 (Lyapunov) *System* (4.14) *with* $A(t) \equiv A$ (*constant matrix*) *is asymptotically stable if and only if there exists an* $n \times n$, *symmetric, positive definite matrix* P *satisfying the equality*

$$A^*P + PA = -\mathbb{1}, \tag{4.50}$$

where, as usual, we denote by A^* *the adjoint* (*transpose*) *of* A, *and by* $\mathbb{1}$ *the identity matrix.*

Proof Assume first that Eq. (4.50) admits a symmetric, positive definite solution P. Consider the function $V : \mathbb{R}^n \to \mathbb{R}$ defined by

$$V(\boldsymbol{x}) = \frac{1}{2}\big(P\boldsymbol{x}, \boldsymbol{x}\big), \quad \forall \boldsymbol{x} \in \mathbb{R}^n. \tag{4.51}$$

Observing that $\operatorname{grad} V(\boldsymbol{x}) = P\boldsymbol{x}$, $\forall \boldsymbol{x} \in \mathbb{R}^n$, we obtain that

$$2\big(\operatorname{grad} V(\boldsymbol{x}), A\boldsymbol{x}\big) = 2(P\boldsymbol{x}, A\boldsymbol{x}) = (A^*P\boldsymbol{x}, \boldsymbol{x}) + (\boldsymbol{x}, PA\boldsymbol{x}) \overset{(4.50)}{=} -\|\boldsymbol{x}\|_{\mathbf{e}}^2. \tag{4.52}$$

On the other hand, since P is positive definite, we deduce (see Lemma A.3) that there exists a $\nu > 0$ such that

$$V(\boldsymbol{x}) \geq \nu\|\boldsymbol{x}\|_{\mathbf{e}}^2, \quad \forall \boldsymbol{x} \in \mathbb{R}^n.$$

The Cauchy–Schwartz inequality implies

$$2V(\boldsymbol{x}) \leq \|P\|_{\mathbf{e}}\|\boldsymbol{x}\|_{\mathbf{e}}^2, \quad \forall \boldsymbol{x} \in \mathbb{R}^n,$$

where

$$\|P\|_{\mathbf{e}} := \sup_{\|\boldsymbol{x}\|_{\mathbf{e}}=1} \|P\boldsymbol{x}\|_{\mathbf{e}}.$$

Thus, V is a Lyapunov function for the differential system

$$\boldsymbol{x}' = A\boldsymbol{x}. \tag{4.53}$$

Since the Euclidean norm $\| - \|_{\mathbf{e}}$ is equivalent to the norm $\| - \|$ (see Lemma A.1), we deduce that the function $\boldsymbol{x} \mapsto (\operatorname{grad} V(\boldsymbol{x}), A\boldsymbol{x})$ is negative definite. Theorem 4.7 now implies that the trivial solution of (4.53) is asymptotically stable.

Conversely, suppose that A is a Hurwitzian matrix. We define a matrix P by the equality

$$P := \int_0^\infty e^{tA^*} e^{tA} dt. \tag{4.54}$$

The above integral is well defined since the real matrix A^* is also Hurwitzian, because its characteristic polynomial coincides with that of A. Theorem 2.13 shows that the entries of both e^{tA} and e^{tA^*} decay exponentially to zero as $t \to \infty$.

Note that $e^{tA^*} = (e^{tA})^*$ so that P is symmetric and

$$(P\boldsymbol{x}, \boldsymbol{x}) = \int_0^\infty \left(e^{tA}\boldsymbol{x}, e^{tA}\boldsymbol{x}\right)dt = \int_0^\infty \|e^{tA}\boldsymbol{x}\|_{\mathrm{e}}^2\, dt, \quad \forall \boldsymbol{x} \in \mathbb{R}^n. \tag{4.55}$$

The above equality implies that P is positive definite. Using (4.54), we get

$$A^*P = \int_0^\infty A^* e^{tA^*} e^{tA} dt = \int_0^\infty \frac{d}{dt}\left(e^{tA^*}\right)e^{tA} dt.$$

Integrating by parts, we obtain

$$A^*P = -\mathbb{1} - \int_0^\infty e^{tA^*}\frac{d}{dt}\left(e^{tA}\right)dt = -\mathbb{1} - \int_0^\infty e^{tA^*} e^{tA} A dt = -\mathbb{1} - PA.$$

Thus P also satisfies equality (4.50). This completes the proof of Theorem 4.8. □

Remark 4.5 Theorem 4.8 can be alternatively rephrased as follows: *the matrix A is Hurwitzian if and only if Eq.* (4.50) *admits a symmetric and positive definite solution P.*

Observe that, if P is a positive definite, symmetric $n \times n$ matrix, then the function

$$(-, -)_P : \mathbb{R}^n \times \mathbb{R}^n \to \mathbb{R}, \quad (\boldsymbol{x}, , \boldsymbol{y})_P := (P\boldsymbol{x}, \boldsymbol{y}),$$

is a Euclidean scalar product on $\mathbb{R}^n$. Theorem 4.8 essentially states that A is Hurwitzian if and only if there exists a Euclidean scalar product $\langle -, - \rangle$ on $\mathbb{R}^n$ such that the resulting symmetric bilinear form

$$Q_A(\boldsymbol{x}, \boldsymbol{y}) = \langle A\boldsymbol{x}, \boldsymbol{y}\rangle + \langle \boldsymbol{x}, A\boldsymbol{y}\rangle$$

is negative definite.

We have to point out that there are no general recipes or techniques for constructing Lyapunov functions. This takes ingenuity and experience. However, the physical intuition behind a given model often suggests candidates for Lyapunov functions. For example, in the case of mechanical, electrical or thermodynamical systems, the energy tends to be a Lyapunov function. In the case of thermodynamical systems, the entropy, taken with the opposite sign, is also a Lyapunov function. We will illustrate this principle in the case of conservative mechanical systems.

Example 4.3 Consider the differential equation

$$x'' + g(x) = 0. \tag{4.56}$$

As explained in Sect. 1.3.5, Eq. (4.56) describes the one-dimensional motion of a unit mass particle under the influence of a force field of the form $-g(x)$, $x(t)$ denoting the position of the particle at time t. (We have discussed the case $g(x) = \sin x$ a bit earlier.) Eq. (4.56) is equivalent to the differential system

$$\begin{aligned} x' &= y, \\ y' &= -g(x). \end{aligned} \tag{4.57}$$

We assume that $g : \mathbb{R} \to \mathbb{R}$ is continuous and satisfies the condition

$$xg(x) > 0, \quad \forall x \in \mathbb{R} \setminus \{0\}.$$

Let G be the antiderivative of g determined by

$$G(x) := \int_0^x g(r)dr, \quad \forall x \in \mathbb{R}.$$

The function G is C^1 and positive away from the origin and thus, according to Remark 4.4, it is positive definite. It is now easy to verify that the function

$$V(x, y) = \frac{1}{2}y^2 + G(x)$$

is conserved along the trajectories of (4.57), which shows that it is a Lyapunov function of this system.

Let us observe that, in the present case, the Lyapunov function V is none other than the total energy of system (4.56). Our assumption on g, which showed that the potential energy G is positive definite, simply states that $x = 0$ is a minimum of G. We have thus recovered through Lyapunov's theorem the well-known principle of Lagrange for conservative systems: *an equilibrium position that is a minimum point for the potential energy is a stable equilibrium.*

However, as we observed earlier, the trivial solution of system (4.57) *is not an asymptotically stable solution*. Indeed, if $(x(t), y(t))$ is a solution of (4.57), then it satisfies the energy conservation law

$$\frac{d}{dt}V\big(x(t), y(t)\big) = 0, \quad \forall t \geq 0,$$

and hence

$$\frac{1}{2}|y(t)|^2 + G\big(x(t)\big) = \frac{1}{2}|y(0)|^2 + G\big(x(0)\big), \quad \forall t \geq 0.$$

If

$$\lim_{t\to\infty} x(t) = \lim_{t\to\infty} y(t) = 0,$$

then

$$\frac{1}{2}|y(0)|^2 + G\big(x(0)\big) = 0.$$

This implies that $x(0) = y(0) = 0$. Thus, the only solution that approaches the trivial solution as $t \to \infty$ is the trivial solution. As a matter of fact, this happens for all physical systems which conserve energy.

For the gradient differential systems, that is, systems of the form

$$\boldsymbol{x}' + \operatorname{grad} f(\boldsymbol{x}) = 0, \tag{4.58}$$

an isolated equilibrium point is asymptotically stable if and only if it is a local minimum of the function $f : \mathbb{R}^n \to \mathbb{R}$. More precisely, we have the following result.

Theorem 4.9 *Let $f : \mathbb{R}^n \to \mathbb{R}$ be a C^2-function. If $\boldsymbol{x}_0$ is an* isolated *critical point of f which is also a local minimum, then the stationary solution $\boldsymbol{x} = \boldsymbol{x}_0$ is an asymptotically stable solution of system* (4.58).

Proof Without loss of generality, we can assume that $\boldsymbol{x}_0 = 0$ and $f(0) = 0$. There exists then a real number $r > 0$ such that

$$f(\boldsymbol{x}) > 0, \quad \operatorname{grad} f(\boldsymbol{x}) \neq 0, \quad \forall 0 < \|\boldsymbol{x}\| < r.$$

This shows that the restriction of f to the disk $\{\|\boldsymbol{x}\| < r\}$ is a Lyapunov function for system (4.58) restricted to this disk. The theorem now follows from Corollary 4.4. □

Remark 4.6 Consider the functions

$$f : \mathbb{R} \to \mathbb{R}, \quad f(x) = \begin{cases} x^5 \left(\sin\left(\frac{1}{x}\right) + 1\right), & x \neq 0, \\ 0, & x = 0, \end{cases}$$

$$F(x) = \int_0^x f(t)dt.$$

Note that F is a C^2-function with a unique minimum at $x = 0$, which is a global minimum. The function F is an antiderivative of f and has infinitely many critical points

$$\frac{1}{-\frac{\pi}{2} + 2n\pi}, \quad n \in \mathbb{Z} \setminus \{0\},$$

which are neither local minima nor local maxima and accumulate at the origin. The trivial solution of the equation

$$x' + F'(x) = 0$$

is stable, but not asymptotically stable. The reason is that the origin is isolated as a local minimum of F, but not isolated as a critical point of F.

Remark 4.7 Theorem 4.3 can be obtained by an alternate proof based on the Lyapunov function method. Indeed, if the matrix A and the perturbation $\boldsymbol{F}$ satisfy the assumptions in Theorem 4.3, then, according to Theorem 4.8, the equation

$$P^*A + AP = -\mathbb{1}$$

admits a symmetric, positive solution P. The function $V(\boldsymbol{x}) = \frac{1}{2}(P\boldsymbol{x}, \boldsymbol{x})$ is then a Lyapunov function for system (4.21). Indeed,

$$\begin{aligned}(A\boldsymbol{x} + \boldsymbol{F}(t, \boldsymbol{x}), \operatorname{grad} V(\boldsymbol{x})) &= (A\boldsymbol{x} + \boldsymbol{F}(t, \boldsymbol{x}), P\boldsymbol{x}) = -\frac{1}{2}\|\boldsymbol{x}\|_{\mathbf{e}}^2 + (\boldsymbol{F}(t, \boldsymbol{x}), P\boldsymbol{x}) \\ &\leq -\frac{1}{2}\|\boldsymbol{x}\|_{\mathbf{e}}^2 + \|\boldsymbol{F}(t, \boldsymbol{x})\|_{\mathbf{e}} \cdot \|P\boldsymbol{x}\|_{\mathbf{e}} \leq -\frac{1}{2}\|\boldsymbol{x}\|_{\mathbf{e}}^2 + CL\|\boldsymbol{x}\|^2.\end{aligned}$$

Thus, if the constant L is sufficiently small, there exists an $\alpha > 0$ such that $\big(A\boldsymbol{x} + \boldsymbol{F}(t, \boldsymbol{x}), \operatorname{grad} V(\boldsymbol{x})\big) \leq -\alpha\|\boldsymbol{x}\|^2$, and so, now Theorem 4.6 implies the stability of the trivial solution.

4.5 Stability of Control Systems

In this section, we will present a few application of the Lyapunov function method to the theory of stability of automatic control systems.

Maxwell's 1868 paper "*On Governors*" and I.A. Vishnegradski's 1877 work on Watt's centrifugal governor represent pioneering contributions to this field. The modern mathematical theory of automatic control systems was put together during the last few decades, and stability theory plays a central part in it.

A *linear control system* is described mathematically by the differential system

$$\boldsymbol{x}' = A\boldsymbol{x} + B\boldsymbol{u}, \tag{4.59}$$

where A is an $n \times m$ matrix, B is an $n \times m$ matrix and $\boldsymbol{u}$ is an $\mathbb{R}^m$-valued function called the *control* or *input*. The function $\boldsymbol{x}$ is called *the state* of the system.

In certain situations, the state of the system is not known directly, but only indirectly through measurements of a certain function $C(\boldsymbol{x})$ of the state. In such cases, we associate to the control system the equation

$$\boldsymbol{y} = C(\boldsymbol{x}). \tag{4.60}$$

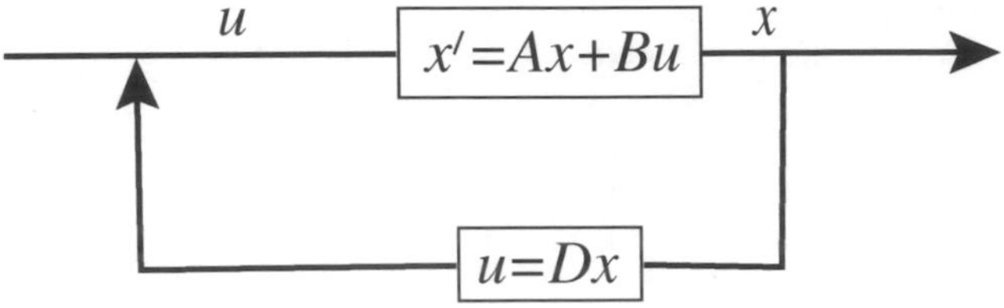

Fig. 4.1 A closed-loop system

The quantity $\boldsymbol{y}$ is called the *output*. The system of Eqs. (4.59) and (4.60) is called an *observed control system*. The control $\boldsymbol{u}$ can also be given as a function of the state $\boldsymbol{x}$ of system (4.59)

$$\boldsymbol{u} = D(\boldsymbol{x}), \tag{4.61}$$

where $D : \mathbb{R}^n \to \mathbb{R}^m$ is a continuous map.

An expression as in (4.61) is called a *feedback synthesis* or *feedback controller* and it arises frequently in the theory of automatic control systems. Putting $\boldsymbol{u}$, given in (4.61), into system (4.59), we obtain the so-called "closed-loop" system

$$\boldsymbol{x}' = A\boldsymbol{x} + BD(\boldsymbol{x}). \tag{4.62}$$

We can schematically describe this closed-loop system as in Fig. 4.1.

Example 4.4 Consider the following simple control system

$$x'(t) = u(t),$$

that could describe the evolution of the volume of fluid in a reservoir where the fluid is added or removed at a rate $u(t)$. Suppose that we want to find the size of the flow $u(t)$ so that, in the long run, the volume will approach a prescribed value x_∞. This goal can be achieved by selecting the flow according to the feedback rule

$$u(t) = \alpha\big(x_\infty - x(t)\big),$$

where α is a positive constant. Indeed, the general solution of the above linear equation is

$$x(t) = x(0)e^{-\alpha t} + x_\infty\big(1 - e^{-\alpha t}\big),$$

and thus

$$\lim_{t\to\infty} x(t) = x_\infty.$$

We say that the feedback synthesis (4.61) *stabilizes* system (4.59) if the corresponding closed-loop system (4.62) is asymptotically stable.

As a matter of fact, the transformation of the original system into a stable or asymptotically stable one is the main motivation in engineering for the design of feedback controls. In practical situations, the feedback control often takes the form of a regulator of the system, making it stable under perturbations, possibly endowing it with other desirable features.

An example available to almost everyone is the classical thermostat that controls the temperature in heated systems. In fact, the main systems in nature and society are stable closed-loop systems.

Example 4.5 Consider the following situation

$$mx'' + \omega^2 x = u(t), \quad t \geq 0, \tag{4.63}$$

that describes the frictionless elastic motion of a particle of mass m acted upon by an external force $u(t)$. If we regard $u(t)$ as a control, then we can rewrite (4.63) in the form (4.59) where the matrix A is given by

$$A := \begin{bmatrix} 0 & 1 \\ -\frac{\omega^2}{m} & 0 \end{bmatrix},$$

and the 2×1 matrix B is given by

$$B := \begin{bmatrix} 0 \\ \frac{1}{m} \end{bmatrix}.$$

As observed in the previous section, the trivial solution is not asymptotically stable. To stabilize this system, we consider a feedback control of the form

$$u(t) = -\alpha x'(t), \tag{4.64}$$

where α is an arbitrary positive constant. The resulting closed-loop system

$$mx'' + \alpha x' + \omega^2 x = 0$$

is obviously asymptotically stable. Concretely, such a system is produced by creating at every moment of time $t > 0$ a resistance force proportional to the velocity of the particle. In other words, we artificially introduce friction into the system.

Consider next the linear control system

$$\boldsymbol{x}' = A\boldsymbol{x} + u(t)\boldsymbol{b}, \quad t \geq 0, \quad \boldsymbol{b} \in \mathbb{R}^n, \tag{4.65}$$

with output σ given by the differential equation

$$\sigma' = (\boldsymbol{c}, \boldsymbol{x}) - \alpha\varphi(\sigma), \tag{4.66}$$

and the nonlinear feedback law

$$u = \varphi(\sigma). \tag{4.67}$$

Above, A is an $n \times n$ Hurwitzian matrix, $\boldsymbol{b}, \boldsymbol{c} \in \mathbb{R}^n$ are control parameters and $\varphi : \mathbb{R} \to \mathbb{R}$ is a C^1-function satisfying

$$\sigma\varphi(\sigma) > 0, \quad \forall \sigma \neq 0, \tag{4.68}$$

$$\lim_{\sigma\to\infty} \int_0^\sigma \varphi(r)dr = \infty. \tag{4.69}$$

We seek conditions that will guarantee the global asymptotic stability of the control system (4.67), that is,

$$\lim_{t\to\infty} \boldsymbol{x}(t) = 0, \quad \lim_{t\to\infty} \sigma(t) = 0, \tag{4.70}$$

for any (unknown) function φ satisfying (4.68) and (4.69).

Such a problem is known as a *Lurie–Postnikov problem*, named after the authors who first investigated it; see e.g. [5, Sect. 6.2] or [9].

To solve this problem, we will seek a Lyapunov function for the system (4.65) and (4.66) of the form

$$V(\boldsymbol{x}, \sigma) = \frac{1}{2}\big(P\boldsymbol{x}, \boldsymbol{x}\big) + \int_0^\sigma \varphi(r)dr, \tag{4.71}$$

where P is a symmetric and positive definite solution of the matrix equation (4.50). From Theorem 4.8 and (4.68) to (4.69), we deduce that the function V is positive definite and

$$\lim_{\|\boldsymbol{x}\|+|\sigma|\to\infty} V(\boldsymbol{x}, \sigma) = \infty.$$

If we denote by $\boldsymbol{f}(\boldsymbol{x}, \sigma)$ the right-hand side of the system (4.65), (4.66) and by $\langle -, - \rangle$ the canonical scalar product on $\mathbb{R}^{n+1}$, we deduce that

$$\begin{aligned}
\langle \operatorname{grad} V(\boldsymbol{x}, \sigma), \boldsymbol{f}(\boldsymbol{x}, \sigma)\rangle &= (A\boldsymbol{x}, P\boldsymbol{x}) + \varphi(\sigma)(P\boldsymbol{x}, \boldsymbol{b}) + \varphi(\sigma)(\boldsymbol{c}, \boldsymbol{x}) - \alpha\varphi(\sigma)^2 \\
&= -\frac{1}{2}\|\boldsymbol{x}\|_{\mathbf{e}}^2 - \alpha\varphi(\sigma)^2 + \varphi(\sigma)(\boldsymbol{x}, P\boldsymbol{b} + \boldsymbol{x}) \\
&\leq -\frac{1}{2}\|\boldsymbol{x}\|_{\mathbf{e}}^2 - \alpha\varphi(\sigma)^2 + \frac{1}{2}\big(\|\boldsymbol{x}\|_{\mathbf{e}}^2 + \varphi(\sigma)^2\|P\boldsymbol{b} + \boldsymbol{c}\|_{\mathbf{e}}^2\big) \\
&= \frac{1}{2}\varphi(\sigma)^2\big(\|P\boldsymbol{b} + \boldsymbol{c}\|_{\mathbf{e}}^2 - 2\alpha\big).
\end{aligned}$$

It follows that the function $\langle \operatorname{grad} V, \boldsymbol{f}\rangle$ is negative definite as soon as $\boldsymbol{b}, \boldsymbol{c}$ satisfy the *controllability inequality*

$$\|P\boldsymbol{b} + \boldsymbol{c}\|_{\mathbf{e}} < \sqrt{2\alpha}.$$

We have thus found a sufficient condition for the asymptotic stability involving the parameters $\alpha, \boldsymbol{b}, \boldsymbol{c}$.

Remark 4.8 The object of the theory of control systems of the form (4.59) involves problems broader than the one we discussed in this section. The main problem of control systems can be loosely described as follows: find the control function $\boldsymbol{u}(t)$ from a class of admissible functions such that the corresponding output $\boldsymbol{x}$ has a prescribed property. In particular, asymptotic stability could be one of those properties.

In certain situations, the control parameter is chosen according to certain optimality conditions, such as, to minimize a functional of the form

$$\int_0^T L\big(\boldsymbol{x}(t), \boldsymbol{u}(t)\big)dt$$

on a collection of functions $\boldsymbol{u}$, where $\boldsymbol{x}$ is the corresponding output. Such a control is called an *optimal control*.

4.6 Stability of Dissipative Systems

We will investigate the asymptotic properties of the system

$$\boldsymbol{x}' = \boldsymbol{f}(\boldsymbol{x}), \tag{4.72}$$

where $\boldsymbol{f} : \mathbb{R}^n \to \mathbb{R}^n$ is a continuous dissipative map (see (2.57))

$$\big(\boldsymbol{f}(\boldsymbol{x}) - \boldsymbol{f}(\boldsymbol{y}), \boldsymbol{x} - \boldsymbol{y}\big) \leq 0, \quad \forall \boldsymbol{x}, \boldsymbol{y} \in \mathbb{R}^n. \tag{4.73}$$

We have seen in Theorem 2.13 that, for any $\boldsymbol{x}_0 \in \mathbb{R}^n$, system (4.72) admits a unique solution $\boldsymbol{x}(t; t_0, \boldsymbol{x}_0)$ satisfying the initial condition $\boldsymbol{x}(t_0) = \boldsymbol{x}_0$ and defined on $[T_0, \infty)$.

If we denote by $S(t) : \mathbb{R}^n \to \mathbb{R}^n$ the semigroup of transformations

$$S(t)\boldsymbol{x}_0 = \boldsymbol{x}(t; 0, \boldsymbol{x}_0), \quad \forall t \geq 0, \quad \boldsymbol{x}_0 \in \mathbb{R}^n,$$

then, obviously,

$$\boldsymbol{x}(t; t_0, \boldsymbol{x}_0) = S(t - t_0)\boldsymbol{x}_0, \quad \forall t \geq t_0, \quad \boldsymbol{x}_0 \in \mathbb{R}^n.$$

Moreover, according to Theorem 2.13, we have

$$\|\boldsymbol{x}(t; t_0, \boldsymbol{x}_0) - \boldsymbol{x}(t; t_0, \boldsymbol{y}_0)\|_{\mathrm{e}} \leq \|\boldsymbol{x}_0 - \boldsymbol{y}_0\|_{\mathrm{e}}, \quad \forall t \geq t_0, \quad \boldsymbol{x}_0, \boldsymbol{y}_0 \in \mathbb{R}^n. \tag{4.74}$$

By the *orbit* or *trajectory* of the system through the point $\boldsymbol{x}_0$, we understand the set

$$\gamma(\boldsymbol{x}_0) := \big\{\boldsymbol{x} \in \mathbb{R}^n;\ \ \boldsymbol{x} = S(t)\boldsymbol{x}_0,\ t \geq 0\big\}.$$

The ω-limit set of the orbit $\gamma(\boldsymbol{x}_0)$ is the set $\omega(\boldsymbol{x}_0)$ of the limit points of $\gamma(\boldsymbol{x}_0)$. More precisely, $\boldsymbol{x} \in \omega(\boldsymbol{x}_0)$ if and only if there exists a sequence $t_k \to \infty$ such that $S(t_k)\boldsymbol{x}_0 \to \boldsymbol{x}$.

Example 4.6 Consider, for example, the solution

$$(x_1, x_2) = (\sin t, \cos t)$$

of the differential system

$$\begin{aligned} x_1' &= x_2, \\ x_2' &= -x_1, \end{aligned}$$

with the initial condition $x_1(0) = 0$, $x_2(0) = 1$. In this case, the ω-limit set of (0, 1) is the unit circle in $\mathbb{R}^2$, centered at the origin.

A sufficient condition for the set $\omega(\boldsymbol{x}_0)$ to be nonempty is the boundedness of the trajectory $\gamma(\boldsymbol{x}_0)$. Let $F \subset \mathbb{R}^n$ be the set of all the stationary solutions of system (4.72), that is,

$$F := \big\{\boldsymbol{x} \in \mathbb{R}^n;\ \ \boldsymbol{f}(\boldsymbol{x}) = 0\big\}. \tag{4.75}$$

Lemma 4.1 *The set F is closed and convex.*

Proof The continuity of $\boldsymbol{f}$ implies that F is closed.

Let $\boldsymbol{x}, \boldsymbol{y} \in F$ and $t \in [0, 1]$. Since $\boldsymbol{f}$ is dissipative, we see that

$$\big(\boldsymbol{f}(\boldsymbol{\xi}), \boldsymbol{\xi} - \boldsymbol{x}\big) \leq 0,\ \ \big(\boldsymbol{f}(\boldsymbol{\xi}), \boldsymbol{\xi} - \boldsymbol{y}\big) \leq 0,\ \ \forall \boldsymbol{\xi} \in \mathbb{R}^n.$$

Hence

$$\big(\boldsymbol{f}(\boldsymbol{\xi}), \boldsymbol{\xi} - \boldsymbol{x}_t\big) \leq 0,\ \ \forall \boldsymbol{\xi} \in \mathbb{R}^n,\ t \in [0, 1],$$

where $\boldsymbol{x}_t = t\boldsymbol{x} + (1-t)\boldsymbol{z}$. Let us now choose $\boldsymbol{\xi} := \varepsilon \boldsymbol{x}_t + (1-\varepsilon)\boldsymbol{z}$, where $0 < \varepsilon < 1$ and $\boldsymbol{z}$ is arbitrary in $\mathbb{R}^n$. By (4.73), we have

$$\big(\boldsymbol{f}(\varepsilon \boldsymbol{x}_t + (1-\varepsilon)\boldsymbol{z}), \boldsymbol{z} - \boldsymbol{x}_t\big) \leq 0,\ \ \forall \boldsymbol{z} \in \mathbb{R}^n,$$

and, letting $\varepsilon \to 1$, we conclude that

$$\big(\boldsymbol{f}(\boldsymbol{x}_t), \boldsymbol{z} - \boldsymbol{x}_t\big) \leq 0,\ \ \forall \boldsymbol{z} \in \mathbb{R}^n.$$

Now, choose $\boldsymbol{z} := \boldsymbol{x}_t + \boldsymbol{f}(\boldsymbol{x}_t)$ to deduce that $\|\boldsymbol{f}(\boldsymbol{x}_t)\|_{\mathbf{e}} \leq 0$, that is, $\boldsymbol{x}_t \in F$, as claimed. □

Proposition 4.2 *We have the equality*

$$F = \{\boldsymbol{x} \in \mathbb{R}^n;\ \ S(t)\boldsymbol{x} = \boldsymbol{x},\ \ \forall t \geq 0\}. \tag{4.76}$$

Proof Clearly, if $\boldsymbol{x}_0 \in F$, then $S(t)\boldsymbol{x}_0 = \boldsymbol{x}_0$, $\forall t \geq 0$, because, according to Theorem 2.13, system (4.72) has a unique solution $\boldsymbol{x}$ satisfying $\boldsymbol{x}(0) = \boldsymbol{x}_0$, namely, the constant solution.

Conversely, let us assume that $S(t)\boldsymbol{x}_0 = \boldsymbol{x}_0$, $\forall t \geq 0$. Then

$$\boldsymbol{f}(\boldsymbol{x}_0) = \frac{d}{dt}\Big|_{t=0} S(t)\boldsymbol{x}_0 = \lim_{t \searrow 0} \frac{1}{t}\big(S(t)\boldsymbol{x}_0 - \boldsymbol{x}_0\big) = 0.$$

This proves that $\boldsymbol{f}(\boldsymbol{x}_0) = 0$, that is, $\boldsymbol{x}_0 \in F$. □

Theorem 4.10 *Suppose that the set F is nonempty. Then the following hold.*

(i) *For any* $\boldsymbol{x}_0 \in \mathbb{R}^n$ *the set* $\omega(\boldsymbol{x}_0)$ *is compact.*
(ii) $S(t)\omega(\boldsymbol{x}_0) \subset \omega(\boldsymbol{x}_0),\ \ \forall t \geq 0,\ \ \forall \boldsymbol{x}_0 \in \mathbb{R}^n$.
(iii) $\|S(t)\boldsymbol{x} - S(t)\boldsymbol{y})\|_{\mathbf{e}} = \|\boldsymbol{x} - \boldsymbol{y}\|_{\mathbf{e}},\ \ \forall \boldsymbol{x}, \boldsymbol{y} \in \omega(\boldsymbol{x}_0),\ t \geq 0$.
(iv) *For any* $\boldsymbol{y} \in F$ *there exists an* $r > 0$, *such that*

$$\omega(\boldsymbol{x}_0) \subset \{\boldsymbol{x} \in \mathbb{R}^n;\ \|\boldsymbol{x} - \boldsymbol{y}\|_{\mathbf{e}} = r\}.$$

If, additionally, $\omega(\boldsymbol{x}_0) \subset F$, *then the limit* $\boldsymbol{x}_\infty = \lim_{t\to\infty} S(t)\boldsymbol{x}_0$ *exists and belongs to the set F.*

Proof (i) Since $F \neq \emptyset$, the trajectory $\gamma(\boldsymbol{x}_0)$ is bounded for any $\boldsymbol{x}_0$. Indeed, if we take the scalar product of the equation

$$\frac{d}{dt}S(t)\boldsymbol{x}_0 = \boldsymbol{f}\big(S(t)\boldsymbol{x}_0\big),\ \ \forall t \geq 0,$$

with $S(t)\boldsymbol{x}_0 - \boldsymbol{y}_0$, where $\boldsymbol{y}_0 \in F$, then we deduce from (4.75) and the dissipativity of $\boldsymbol{f}$ that

$$\frac{d}{dt}\big\|S(t)\boldsymbol{x}_0 - \boldsymbol{y}_0\big\|_{\mathbf{e}}^2 \leq 0,\ \ \forall t \geq 0.$$

Hence

$$\big\|\, S(t)\boldsymbol{x}_0 - \boldsymbol{y}_0 \,\big\|_{\mathbf{e}} \leq \|\boldsymbol{x}_0 - \boldsymbol{y}_0\|_{\mathbf{e}},\ \ \forall t \geq 0,$$

so that

$$\|S(t)\boldsymbol{x}_0\|_{\mathbf{e}} \leq \|\boldsymbol{y}_0\|_{\mathbf{e}} + \|\boldsymbol{x}_0 - \boldsymbol{y}_0\|_{\mathbf{e}},\ \ \forall t \geq 0.$$

Since the trajectory $\gamma(\boldsymbol{x}_0)$ is bounded, we deduce that the ω-limit set $\omega(\boldsymbol{x}_0)$ is bounded as well. Let us show that the set $\omega(\boldsymbol{x}_0)$ is also closed.

Consider a sequence $\{\boldsymbol{x}_j\} \subset \omega(\boldsymbol{x}_0)$ that converges to $\boldsymbol{\xi} \in \mathbb{R}^n$. For any $\boldsymbol{x}_j$ there exists a sequence of positive real numbers $\{t_{j,k}\}_{k\geq 1}$ such that

$$\lim_{k\to\infty} t_{j,k} = \infty \text{ and } \boldsymbol{x}_j = \lim_{k\to\infty} S(t_{j,k})\boldsymbol{x}_0.$$

Hence, for any j and any $\varepsilon > 0$, there exists a $K = K(\varepsilon, j)$ such that

$$\|S(t_{j,k})\boldsymbol{x}_0 - \boldsymbol{x}_j\|_{\mathbf{e}} \leq \varepsilon, \quad \forall k \geq K(\varepsilon, j).$$

Thus

$$\|S(t_{j,k})\boldsymbol{x}_0 - \boldsymbol{\xi}\|_{\mathbf{e}} \leq \varepsilon + \|\boldsymbol{x}_j - \boldsymbol{\xi}\|_{\mathbf{e}}, \quad \forall k > K(\varepsilon, j).$$

Since $\boldsymbol{x}_j \to \boldsymbol{\xi}$ as $t \to \infty$, we deduce that there exists a $j(\varepsilon) > 0$ such that

$$\|\boldsymbol{x}_j - \boldsymbol{\xi}\|_{\mathbf{e}} \leq \varepsilon, \quad \forall j \geq j(\varepsilon).$$

In particular, for $j = j(\varepsilon)$ and $k := k(\varepsilon) = K(\varepsilon, j(\varepsilon))$, we have

$$\|S(t_{j(\varepsilon),k(\varepsilon)})\boldsymbol{x}_0 - \boldsymbol{\xi}\|_{\mathbf{e}} \leq 2\varepsilon.$$

This proves that $\boldsymbol{\xi}$ is a limit point of the sequence $\{S((t_{j,k})\boldsymbol{x}_0\}_{j,k\geq 1}$, that is, $\boldsymbol{\xi} \in \omega(\boldsymbol{x}_0)$.

(ii) Let $\boldsymbol{\xi} \in \omega(\boldsymbol{x}_0)$. There exists a sequence $t_k \to \infty$ such that

$$\lim_{k\to\infty} S(t_k)\boldsymbol{x}_0 = \boldsymbol{\xi}.$$

According to Theorem 2.13, the map $S(t) : \mathbb{R}^n \to \mathbb{R}^n$ is continuous for any $t \geq 0$ and thus

$$\lim_{k\to\infty} S(t + t_k)\boldsymbol{x}_0 = \lim_{k\to\infty} S(t)S(t_k)\boldsymbol{x}_0 = S(t)\boldsymbol{\xi}, \quad \forall t \geq 0.$$

Hence $S(t)\boldsymbol{\xi} \in \omega(\boldsymbol{x}_0)$, so that $S(t)\omega(\boldsymbol{x}_0) \subset \omega(\boldsymbol{x}_0)$.

(iii) Let us first prove that, if $\boldsymbol{x} \in \omega(\boldsymbol{x}_0)$, then there exist $s_k \to \infty$ such that $S(s_k)\boldsymbol{x} \to \boldsymbol{x}$ as $k \to \infty$.

Indeed, there exist $t_k \to \infty$ such that

$$t_{k+1} - t_k \to \infty, \quad S(t_k)\boldsymbol{x}_0 = \boldsymbol{x}.$$

If we set $s_k := t_{k+1} - t_k$, then we get $S(s_k)S(t_k) = S(t_{k+1})$ and

$$\begin{aligned}\|S(s_k)\boldsymbol{x} - \boldsymbol{x}\|_{\mathbf{e}} &\leq \|S(s_k)\boldsymbol{x} - S(s_k)S(t_k)\boldsymbol{x}_0\|_{\mathbf{e}} + \|S(t_{k+1})\boldsymbol{x}_0 - \boldsymbol{x}\|_{\mathbf{e}} \\ &\overset{(2.61)}{\leq} \|\boldsymbol{x} - S(t_k)\boldsymbol{x}_0\|_{\mathbf{e}} + \|S(t_{k+1})\boldsymbol{x}_0 - \boldsymbol{x}\|_{\mathbf{e}} \to 0.\end{aligned}$$

Thus $\omega(\boldsymbol{x}_0) \subset \omega(\boldsymbol{x})$, $\forall \boldsymbol{x} \in \omega(\boldsymbol{x}_0)$. The opposite inclusion is obvious and thus

$$\omega(\boldsymbol{x}_0) = \omega(\boldsymbol{x}), \quad \forall \boldsymbol{x} \in \omega(\boldsymbol{x}_0).$$

Let $\boldsymbol{x}, \boldsymbol{y} \in \omega(\boldsymbol{x}_0)$. Then $\boldsymbol{y} \in \omega(\boldsymbol{x})$ and there exist sequences $s_k, \tau_k \to \infty$ such that

$$\boldsymbol{x} = \lim_{k\to\infty} S(s_k)\boldsymbol{x}, \quad \boldsymbol{y} = \lim_{k\to\infty} S(\tau_k)\boldsymbol{x}.$$

From the semigroup and the contraction properties of the family $S(t)$, we deduce the following string of inequalities

$$\begin{aligned}\|S(\tau_k)\boldsymbol{y} - \boldsymbol{y}\|_{\mathbf{e}} &\le \|S(s_k)\boldsymbol{y} - S(\tau_k + s_k)\boldsymbol{x}\|_{\mathbf{e}} + \|S(\tau_k + s_k)\boldsymbol{x} - S(\tau_k)\boldsymbol{x}\|_{\mathbf{e}} \\ &+ \|S(\tau_k)\boldsymbol{x} - \boldsymbol{y}\|_{\mathbf{e}} \le 2\|S(\tau_k)\boldsymbol{x} - \boldsymbol{y}\|_{\mathbf{e}} + \|S(s_k)\boldsymbol{x} - \boldsymbol{x}\|_{\mathbf{e}} \to 0.\end{aligned}$$

Hence

$$\boldsymbol{x} = \lim_{k\to\infty} S(s_k)\boldsymbol{x}, \quad \boldsymbol{y} = \lim_{k\to\infty} S(s_k)\boldsymbol{y},$$

and we deduce that

$$\begin{aligned}\|\boldsymbol{x} - \boldsymbol{y}\|_{\mathbf{e}} &= \lim_{k\to\infty} \|S(s_k)\boldsymbol{x} - S(s_k)\boldsymbol{y}\|_{\mathbf{e}} \\ &= \lim_{k\to\infty} \|S(s_k - t)S(t)\boldsymbol{x} - S(s_k - t)S(t)\boldsymbol{y}\|_{\mathbf{e}} \\ &\le \|S(t)\boldsymbol{x} - S(t)\boldsymbol{y}\|_{\mathbf{e}}.\end{aligned} \tag{4.77}$$

By Theorem 2.13, the map $S(t)$ is a contraction so that

$$\|S(t)\boldsymbol{x} - S(t)\boldsymbol{y}\|_{\mathbf{e}} \le \|\boldsymbol{x} - \boldsymbol{y}\|_{\mathbf{e}}.$$

This proves the equality (iii).

(iv) Let $\boldsymbol{y} \in F$. Scalarly multiplying the equation

$$\frac{d}{dt}\big(S(t)\boldsymbol{x}_0 - \boldsymbol{y}\big) = \boldsymbol{f}\big(S(t)\boldsymbol{x}_0\big), \quad t \ge 0,$$

by $S(t)\boldsymbol{x}_0 - \boldsymbol{y}$ and using the dissipativity condition (4.73), we find that

$$\frac{d}{dt}\| S(t)\boldsymbol{x}_0 - \boldsymbol{y}\|_{\mathbf{e}}^2 \le 0, \quad \forall t \ge 0.$$

Thus the limit

$$r := \lim_{t\to\infty} \|S(t)\boldsymbol{x}_0 - \boldsymbol{y}\|_{\mathbf{e}}$$

exists. From the definition of the set $\omega(\boldsymbol{x}_0)$, we get that

$$\|\boldsymbol{x} - \boldsymbol{y}\|_{\mathbf{e}} = r, \quad \forall \boldsymbol{x} \in \omega(\boldsymbol{x}_0).$$

Suppose now that $\omega(\boldsymbol{x}_0) \subset F$. Let $\boldsymbol{y}$ be an arbitrary point in $\omega(\boldsymbol{x}_0)$. From the above discussion, it follows that there exists an $r \geq 0$ such that

$$\|\boldsymbol{x} - \boldsymbol{y}\|_{\mathrm{e}} = r, \quad \forall \boldsymbol{x} \in \omega(\boldsymbol{x}_0).$$

Since $\boldsymbol{y} \in \omega(\boldsymbol{x}_0)$, it follows that $r = 0$ and thus $\omega(\boldsymbol{x}_0)$ is reduced to a single point

$$\boldsymbol{x}_\infty = \lim_{t\to\infty} S(t)\boldsymbol{x}_0.$$

This completes the proof of Theorem 4.10. □

Theorem 4.10 is a very valuable tool for investigating asymptotic stability in certain situations not covered by the Lyapunov function method, or the first approximation method.

Example 4.7 Consider the dissipative system (4.72) satisfying the additional conditions

$$\boldsymbol{f}(0) = 0, \quad \big(\boldsymbol{f}(\boldsymbol{x}), \boldsymbol{x}\,\big) < 0, \quad \forall \boldsymbol{x} \neq 0. \tag{4.78}$$

We want to show that, under these assumptions, the trivial solution of system (4.72) is globally asymptotically stable, that is,

$$\lim_{t\to\infty} \boldsymbol{x}(t; 0, \boldsymbol{x}_0) = 0, \quad \forall \boldsymbol{x}_0 \in \mathbb{R}^n. \tag{4.79}$$

Taking into account (4.74), we can rewrite the above equality as

$$\lim_{t\to\infty} S(t)\boldsymbol{x}_0 = 0, \quad \forall \boldsymbol{x}_0 \in \mathbb{R}^n.$$

In view of Theorem 4.10, it suffices to show that

$$\omega(\boldsymbol{x}_0) = F = \{0\}. \tag{4.80}$$

Assumption (4.78) implies that $F = \{0\}$. Let us first observe that (4.78) also implies that any trajectory $\gamma(\boldsymbol{x}_0)$ of system (4.72) is bounded.

Indeed, if $\boldsymbol{x}(t)$ is an arbitrary solution of (4.72), we have

$$\big(\boldsymbol{x}(t), \boldsymbol{x}'(t)\,\big) = \frac{1}{2}\frac{d}{dt}\|\,\boldsymbol{x}(t)\,\|_{\mathbf{e}}^2 = \big(\boldsymbol{f}(\boldsymbol{x}(t)), \boldsymbol{x}(t)\,\big) \leq 0.$$

Hence

$$\|\boldsymbol{x}(t)\|_{\mathbf{e}} \leq \|\boldsymbol{x}(0)\|_{\mathbf{e}}, \quad \forall t \geq 0.$$

This shows that $\gamma(\boldsymbol{x}_0)$ is bounded and thus $\omega(\boldsymbol{x}_0) \neq \emptyset$.

Fix an arbitrary $\boldsymbol{y}_0 \in \omega(\boldsymbol{x}_0)$ and set $\boldsymbol{y}(t) := S(t)\boldsymbol{x}_0$. We have

$$\boldsymbol{y}'(t) = \boldsymbol{f}\big(\boldsymbol{y}(t)\big), \quad \forall t \geq 0,$$

and, arguing as above, we deduce that

$$\frac{1}{2}\frac{d}{dt}\|\boldsymbol{y}(t)\|_{\mathbf{e}}^2 = \big(\boldsymbol{f}(\boldsymbol{y}(t)), \boldsymbol{y}(t)\big), \quad \forall t \geq 0. \tag{4.81}$$

Integrating the above equality, we deduce that

$$\frac{1}{2}\|\boldsymbol{y}(t)\|_{\mathbf{e}}^2 = \int_0^t \big(\boldsymbol{f}(\boldsymbol{y}(s)), \boldsymbol{y}(s)\big)ds + \frac{1}{2}\|\boldsymbol{y}_0\|_{\mathbf{e}}^2, \quad \forall t \geq 0.$$

Since $\big(\boldsymbol{f}(\boldsymbol{y}(s)), \boldsymbol{y}(s)\big) \leq 0$, it follows from the above equality that the integral

$$\int_0^\infty \big(\boldsymbol{f}(\boldsymbol{y}(s)), \boldsymbol{y}(s)\big)ds$$

is convergent and thus there exists a sequence $s_k \to \infty$ such that $\boldsymbol{y}(s_k) \to \boldsymbol{y}_\infty$ and

$$0 = \lim_{k\to\infty} \big(\boldsymbol{f}(\boldsymbol{y}(s_k)), \boldsymbol{y}(s_k)\big) = \big(\boldsymbol{f}(\boldsymbol{y}_\infty), \boldsymbol{y}_\infty\big).$$

Thus $\boldsymbol{y}_\infty \in \omega(\boldsymbol{x}_0)$ and $\big(\boldsymbol{f}(\boldsymbol{y}_\infty), \boldsymbol{y}_\infty\big) = 0$. Assumption (4.78) implies that $\boldsymbol{y}_\infty = 0$ and thus $0 \subset \omega(\boldsymbol{x}_0)$. Theorem 4.10(iii) implies that

$$\|\boldsymbol{y}(t)\|_{\mathbf{e}} = \|\boldsymbol{y}_0\|_{\mathbf{e}}, \quad \forall t \geq 0.$$

Using this in (4.81), we deduce that

$$\big(\boldsymbol{f}(\boldsymbol{y}(t)), \boldsymbol{y}(t)\big) = 0, \quad \forall t \geq 0$$

and thus $\boldsymbol{y}(t) = 0$, $\forall t \geq 0$. Hence, $\boldsymbol{y}_0 \in F = \{0\}$, $\forall \boldsymbol{y}_0 \in \omega(\boldsymbol{x}_0)$. This proves (4.80).

Example 4.8 Consider now system (4.72) in the special case $n = 2$. Suppose that $\boldsymbol{f}(0) = 0$ and $F \supsetneqq \{0\}$. Since F is convex, it contains at least one line segment; see Fig. 4.2.

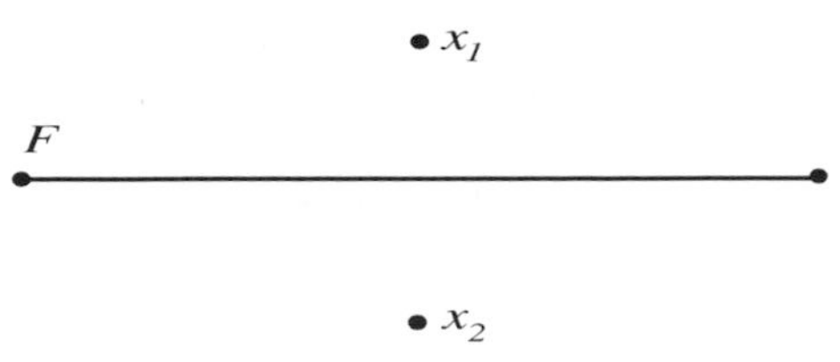

Fig. 4.2 Asymptotic behavior of a two-dimensional dissipative system

On the other hand, according to Theorem 4.10(iv), the set $\omega(\boldsymbol{x}_0)$ is situated on a circle centered at some arbitrary point of the set F. Thus, $\omega(\boldsymbol{x}_0)$ contains at most two points $\boldsymbol{x}_1, \boldsymbol{x}_2$ as in Fig. 4.2. Since

$$S(t)\omega(\boldsymbol{x}_0) \subset \omega(\boldsymbol{x}_0) \text{ and } \lim_{t\to 0} S(t)\boldsymbol{x} = \boldsymbol{x},$$

it follows that $S(t)\boldsymbol{x}_1 = \boldsymbol{x}_1$, $S(t)\boldsymbol{x}_2 = \boldsymbol{x}_2$, $\forall t \geq 0$. Thus, $\boldsymbol{x}_1, \boldsymbol{x}_2 \in F$, which implies that $\boldsymbol{x}_1 = \boldsymbol{x}_2 = \boldsymbol{x}_\infty$ and, therefore,

$$\lim_{t\to\infty} \boldsymbol{x}(t; 0, \boldsymbol{x}_0) = \boldsymbol{x}_\infty.$$

If $F = \{0\}$, then $\omega(\boldsymbol{x}_0)$ is contained in a circle of radius $r \geq 0$ centered at the origin. If $r = 0$, then $\omega(\boldsymbol{x}_0) = \{0\}$.

Remark 4.9 The method of ω-limit sets represents more recent contributions to the development of stability theory, due mainly to G.D. Birkhoff, J.P. LaSalle, C. Dafermos and others.

Problems

4.1 Prove that the matrix A is Hurwitzian if and only if any solution $\boldsymbol{x}(t)$ of the system $\boldsymbol{x}' = A\boldsymbol{x}$ is absolutely integrable on $[0, \infty)$, that is,

$$\int_0^\infty \|\boldsymbol{x}(t)\|dt < \infty. \tag{4.82}$$

Hint. Let $\boldsymbol{x}(t, \boldsymbol{x}_0) = e^{tA}\boldsymbol{x}_0$ and $\varphi(\boldsymbol{x}_0) = \int_0^\infty \|\boldsymbol{x}(t, \boldsymbol{x}_0)\|dt$. By (4.82), it follows that

$$\sup_{t\geq 0} \|\boldsymbol{x}(t, \boldsymbol{x}_0)\| < \infty$$

and, taking into account that $\boldsymbol{x}(t+s, \boldsymbol{x}_0) = \boldsymbol{x}(t, \boldsymbol{x}(s, \boldsymbol{x}_0))$, $\forall t, s \geq 0$, we get

$$\varphi(\boldsymbol{x}(t, \boldsymbol{x}_0)) = \int_t^\infty \|\boldsymbol{x}(s, \boldsymbol{x}_0)\|ds \to 0 \text{ as } t \to \infty.$$

Hence, any limit point $\boldsymbol{x}_0 = \lim_{t_n\to\infty} \boldsymbol{x}(t_n, \boldsymbol{x}_0)$ satisfies $\varphi(\boldsymbol{x}_\infty) = 0$, that is, $\boldsymbol{x}_\infty = 0$.

4.2 Find the stationary solutions of the equation

$$x'' + ax' + 2bx + 3x^2 = 0, \quad a, b > 0 \tag{4.83}$$

and then investigate their stability.
Hint. Stationary solutions are $x_1(t) = 0,\ x_2(t) = -\frac{2b}{3}\cdot$

4.3 Find the stationary solutions of the systems below and then investigate their stability using the first-order-approximation method.

$$x_1' = \sin(x_+ x_2), \quad x_2' = e^{x_1} - 1, \tag{4.84}$$

$$x_1' = 1 - x_1 x_2, \quad x_2' = x_1 - x_2, \tag{4.85}$$

$$x_1' = x_1 x_2 + x_2 \cos(x_1^2 + x_2^2), \quad x_2' = -x_1^2 + x_2 \cos(x_1^2 + x_2^2). \tag{4.86}$$

4.4 Investigate the stability of the stationary solutions of the Lotka–Volterra system

$$\begin{aligned} x_1' &= x_1(1 - x_2), \\ x_2' &= x_2(x_1 - 1). \end{aligned} \tag{4.87}$$

Hint. System (4.87) has two stationary solutions $(0, 0)$ and $(1, 1)$. The first solution is not stable. Translating to the origin the solution $(1, 1)$, we obtain that the function $V(y_1, y_2) = \left(y_1 + y_2 + \ln \frac{1+y_1}{1+y_2}\right)^2$ is a Lyapunov function of the system thus obtained.

4.5 The differential system

$$\begin{aligned} m\varphi'' &= mn^2\omega^2 \sin\varphi \cos\varphi - mg \sin\varphi - b\varphi' \\ \lambda\omega' &= k \cos\varphi - F \end{aligned} \tag{4.88}$$

describes the dynamics of J. Watt's centrifugal governor; see L.C. Pontryagin [17]. Use the first-order-approximation method to study the asymptotic stability of the stationary solution $\varphi = \varphi_0, \omega = \omega_0$, to (4.88).

4.6 Let $H : \mathbb{R}^n \times \mathbb{R}^n \to \mathbb{R}$ be a positive definite C^1-function that is zero at $(0, 0)$. Prove that the trivial solution of the Hamiltonian system

$$\boldsymbol{x}' = \frac{\partial H}{\partial \boldsymbol{p}}(\boldsymbol{x}, \boldsymbol{p}), \quad \boldsymbol{p}' = -\frac{\partial H}{\partial \boldsymbol{x}}(\boldsymbol{x}, \boldsymbol{p}), \tag{4.89}$$

is stable but not asymptotically stable.
Hint. The Hamiltonian function H is a Lyapunov function of system (4.89).

4.7 Prove that the null solution of the system

$$x' = y, \quad y' = -\sin x + y,$$

is not stable.

4.8 Investigate the stability of the stationary solutions of the system

$$x' = y - f(x), \quad y' = -x, \tag{4.90}$$

where $f : \mathbb{R} \to \mathbb{R}$ is a C^1 function.

Hint. System (4.90) is equivalent to the *Liénard equation*

$$x'' + x'f(x) + x = 0$$

that arises in the theory of electrical circuits.

In the special case $f(x) = x^3 - x$, the equation is known as the *Van der Pol equation*. The first-order-approximation method shows that the stationary solution $(0, f(0))$ is stable if $f'(0) > 0$. One can reach the same conclusion from Theorem 4.6 by constructing a Lyapunov function of the form

$$V(x, y) = \alpha x^2 + \beta y^2, \ \alpha, \beta > 0.$$

4.9 Use the Lyapunov function method to prove that the trivial solution of the damped pendulum equation

$$x'' + bx' + \sin x = 0, \tag{4.91}$$

where $b > 0$, is asymptotically stable.
Hint. The equivalent system

$$x' = y, \quad y' = -by - \sin x,$$

admits Lyapunov functions of the form

$$V(x, y) = \alpha y^2 + \beta(1 - \cos x) + \gamma xy, \tag{4.92}$$

for some suitable positive constants α, β, γ.

4.10 Let A be a real $n \times n$ matrix that is nonpositive, that is,

$$(A\boldsymbol{x}, \boldsymbol{x}) \leq 0, \quad \forall \boldsymbol{x} \in \mathbb{R}^n, \tag{4.93}$$

and let B be a real $n \times m$ matrix such that

$$\text{rank}\,[B, AB, A^2B, \ldots, A^{n-1}B] = n. \tag{4.94}$$

Prove that the matrix $A - BB^*$ is Hurwitzian.
Hint. As shown in Problem 3.23, assumption (4.94) implies that

$$B^* e^{tA^*}\boldsymbol{x} = 0, \quad \forall t \geq 0 \quad \text{if and only if } \boldsymbol{x} = 0. \tag{4.95}$$

It suffices to show that the matrix $A^* - BB^*$ is Hurtwitzian. To this end, scalarly multiply the system

$$\boldsymbol{y}' = (A^* - BB^*)\boldsymbol{y} \tag{4.96}$$

by $\boldsymbol{y}$ and then integrate over $[0, \infty)$. We deduce that

$$\int_0^\infty \|B^*\boldsymbol{y}(t)\|_{\mathbf{e}}^2 dt < \infty$$

for any solution of system (4.96). Assumption (4.95) then implies that

$$V(\boldsymbol{x}) = \int_0^\infty \|B^* e^{t(A-BB^*)}\boldsymbol{x}\|_{\mathbf{e}}^2 dt$$

is a Lyapunov function for system (4.96) and $V\big(\boldsymbol{y}(t)\big) \to 0$ as $t \to \infty$.

Remark 4.10 The preceding problem shows that, under assumptions (4.93) and (4.94), the system $\boldsymbol{x}' = A\boldsymbol{x} + B\boldsymbol{u}$ can be stabilized using the feedback controller

$$\boldsymbol{u} = -B^*\boldsymbol{x}. \tag{4.97}$$

4.11 Prove, using a suitable Lyapunov function, that the trivial solution of the system

$$x_1' = -2x_1 + 5x_2 + x_2^2, \;\; x_2' = -4x_1 = 2x_2 + x_1^2$$

is asymptotically stable.

4.12 Using the Lyapunov function method, investigate the stability of the trivial solution of the ODE

$$x'' + a(t)x' + b(t)x = 0.$$

Hint. Seek a Lyapunov function of the form

$$V(x_1, x_2) = x_1^2 + \frac{x_2^2}{b(t)}.$$

4.13 Consider the differential system

$$\boldsymbol{x}' = \boldsymbol{f}(\boldsymbol{x}) + \sum_{i=1}^N u_i B_i(\boldsymbol{x}), \;\; \boldsymbol{x} \in \mathbb{R}^n, \tag{4.98}$$

where u_i, $i = 1, \dots, n$, are real parameters and $\boldsymbol{f} : \mathbb{R}^n \to \mathbb{R}^n$, $B_i : \mathbb{R}^n \to \mathbb{R}^n$, $i = 1, \dots, N$, are locally Lipschitz functions. We assume that there exists a positive definite function V such that

$$\big(\boldsymbol{f}(\boldsymbol{x}), \text{grad}\, V(\boldsymbol{x})\big) \le 0, \;\; \forall \boldsymbol{x} \in \mathbb{R}^n, \tag{4.99}$$

and the functions $\big(f(x), \operatorname{grad} V(x)\big)$, $\big(B_i(x), \operatorname{grad} V(x)\big)$, $i = 1, \ldots, N$, are not simultaneously zero on $\mathbb{R}^n$. Prove that the feedback controller

$$u_i := -\big(B_i(x), \operatorname{grad} V(x)\big),$$

stabilizes system (4.98).
Hint. Verify that V is a Lyapunov function for the differential system

$$x' = f(x) - \sum_{i=1}^{N} \big(B_i(x), \operatorname{grad} V(x)\big) B_i(x).$$

4.14 The differential system

$$N' = -\frac{\alpha}{\ell}(T - T_0)N, \quad mCT' = -N - N_0, \tag{4.100}$$

is a simplified model for the behavior of a nuclear reactor, $N = N(t)$ denoting the power of the reactor at time t, $T = T(t)$ is the temperature, ℓ is the lifetime of the neutrons, m is the mass of radioactive material and α, C are some positive parameters. Investigate the stability of the stationary solution $N = N_0 > 0$, $T = T_0$.
Hint. Making the change of variables

$$x_1 = \ln \frac{N}{N_0}, \quad x_2 = T - T_0,$$

the problem can be reduced to investigating the stability of the null solution of the differential system

$$x_1' = -\mu x_2, \quad x_2' = e^{x_1} - 1, \quad \mu := \frac{\alpha m C}{N_0 \ell},$$

which satisfies the assumptions of Theorem 4.7 for a Lyapunov function of the form

$$V(x_1, x_2) = \frac{\mu}{2} x_2^2 + \int_0^{x_1} (e^s - 1) ds.$$

4.15 Consider the control system

$$x' + ax = u, \quad t \geq 0, \tag{4.101}$$

with the feedback synthesis $u = -\frac{\rho x}{|x|}$, where a, ρ are positive constants. Prove that the solutions of the system that are not zero at $t = 0$ will reach the value zero in a finite amount of time. Find a physical model for this system.

Hint. By multiplying (4.101) by $\frac{x}{|x|}$, it follows that

$$\frac{d}{dt}\,|x(t)| + a|x(t)| + \rho = 0 \text{ on } [t \geq 0;\ x(t) \neq 0],$$

which implies that $x(t) = 0$ for $t \geq T = \frac{1}{a}\,\log\left(\frac{a|x(0)|}{\rho} + 1\right)$.

4.16 Consider the system

$$\boldsymbol{x}' + \text{grad}f(\boldsymbol{x}) = 0, \tag{4.102}$$

where $f : \mathbb{R}^n \to \mathbb{R}$ is a C^2-function such that, for any $\lambda \in \mathbb{R}$, the set

$$\left\{\boldsymbol{x} \in \mathbb{R}^n;\ f(x) \leq \lambda\right\}$$

is bounded and the equation $\text{grad}f(\boldsymbol{x}) = 0$ has finitely many solutions $\boldsymbol{x}_1, \dots, \boldsymbol{x}_m$. Prove that any solution $\boldsymbol{x} = \varphi(t)$ of (4.102) is defined on the entire semi-axis $[0, \infty)$ and $\lim_{t\to\infty} \varphi(t)$ exists and is equal to one of the stationary points $\boldsymbol{x}_1, \dots, \boldsymbol{x}_m$.
Hint. Scalarly multiplying the system (4.102) by $\varphi'(t)$, we deduce that

$$\frac{1}{2}\|\varphi'(t)\|_{\mathbf{e}}^2 + \frac{d}{dt}f\big(\varphi(t)\big) = 0,$$

on the maximal existence interval $[0, T[$. Thus

$$\frac{1}{2}\int_0^t \|\varphi'(s)\|_{\mathbf{e}}^2\,ds + f\big(\varphi(t)\big) = f\big(\varphi(0)\big), \quad \forall t \in [0, T[.$$

Invoking Theorem 3.10, we deduce from the above inequality that $T = \infty$, $\varphi(t)$ is bounded on $[0, \infty)$ and $\lim_{t\to\infty} \varphi'(t) = 0$. Then, one applies Theorem 4.10.

Chapter 5
Prime Integrals and First-Order Partial Differential Equations

In this chapter, we will investigate the concept of a *prime integral* of a system of ODEs and some of its consequences in the theory of first-order partial differential equations. An important part of this chapter is devoted to the study of the Cauchy problem for such partial differential equations. These play an important role in mathematical physics, mechanics and the calculus of variations. The treatment of such problems is essentially of a geometric nature and it is based on properties of systems of ODEs.

5.1 Prime Integrals of Autonomous Differential Systems

Consider the autonomous system

$$\boldsymbol{x}' = \boldsymbol{f}(\boldsymbol{x}), \quad \boldsymbol{x} = (x_1, \dots, x_n), \tag{5.1}$$

where $\boldsymbol{f} : D \to \mathbb{R}^n$ is a C^1-map on an open subset $D \subset \mathbb{R}^n$. We begin by defining the concept of a *prime integral*.

Definition 5.1 The scalar C^1-function $U(\boldsymbol{x}) = U(x_1, \dots, x_n)$ defined on an open set $D_0 \subset D$ is called a *prime integral* of system (5.1) if it is not identically constant, but $U\big(\varphi(t)\big)$ is constant for any trajectory $\boldsymbol{x} = \varphi(t)$ of system (5.1) that stays in D_0.

Theorem 5.1 *The C^1-function U on D_0 is a prime integral of system (5.1) if and only if*

$$\big(\mathit{grad}\, U(\boldsymbol{x}), \boldsymbol{f}(\boldsymbol{x})\big) = 0, \quad \forall \boldsymbol{x} \in D_0. \tag{5.2}$$

Proof If U is a prime integral, then $U(\varphi(t))$ is constant for any solution $\varphi(t)$ of system (5.1). Thus

V. Barbu, *Differential Equations*, Springer Undergraduate Mathematics Series,
DOI 10.1007/978-3-319-45261-6_5

$$0 = \frac{d}{dt}U(\varphi(t)) = \sum_{i=1}^{n} \frac{\partial U}{\partial x_i}(\varphi(t))\boldsymbol{f}_i(\varphi(t)) = (grad\, U(\varphi(t)), \boldsymbol{f}(\varphi(t))). \quad (5.3)$$

Since any point of D_0 is contained in a trajectory of (5.1), we deduce that (5.2) holds on D_0. Conversely, (5.2) implies (5.3) which in turn implies that $U\big(\varphi(t)\big)$ is constant for any solution $\varphi(t)$ of system (5.1). □

To investigate the existence of prime integrals, we need to introduce several concepts.

A point $\boldsymbol{a} \in \mathbb{R}^n$ is called a *critical point* of system (5.1) if $\boldsymbol{f}(\boldsymbol{a}) = 0$. The point is called *regular* if $\boldsymbol{f}(\boldsymbol{a}) \neq 0$.

The C^1-functions $U_1, \ldots, U_k$, $k \leq n$, are called *independent* in a neighborhood of $\boldsymbol{a} \in \mathbb{R}^n$ if the *Jacobian matrix*

$$\left[\frac{\partial U_i}{\partial x_j}(\boldsymbol{a})\right]_{\substack{1\leq i\leq k,\\ 1\leq j\leq n}} \quad (5.4)$$

has rank k.

Equivalently, this means that the vectors $grad\, U_1(\boldsymbol{a}), \ldots, grad\, U_k(\boldsymbol{a})$ are linearly independent.

Theorem 5.2 *In a neighborhood of a regular point $\boldsymbol{a} \in \mathbb{R}^n$ of system (5.1), there exist exactly $(n-1)$-independent prime integrals.*

Proof Let $\boldsymbol{a} \in \mathbb{R}^n$ such that $\boldsymbol{f}(\boldsymbol{a}) \neq 0$. Assume that the n-th component of the vector $\boldsymbol{f}(a)$ is not zero

$$\boldsymbol{f}_n(\boldsymbol{a}) \neq 0. \quad (5.5)$$

We first prove that there exist *at most* $(n-1)$-independent prime integrals in the neighborhood of $\boldsymbol{a}$.

We argue by contradiction and assume that there exist n independent prime integrals $U_1, \ldots, U_n$. From (5.2), we obtain that

$$\begin{aligned} \frac{\partial U_1}{\partial x_1}(\boldsymbol{a})\boldsymbol{f}_1(\boldsymbol{a}) + \cdots + \frac{\partial U_1}{\partial x_n}(\boldsymbol{a})\boldsymbol{f}_n(\boldsymbol{a}) &= 0 \\ \vdots \qquad\qquad\qquad\qquad \vdots\ &\ \vdots \\ \frac{\partial U_n}{\partial x_1}(\boldsymbol{a})\boldsymbol{f}_1(\boldsymbol{a}) + \cdots + \frac{\partial U_n}{\partial x_n}(\boldsymbol{a})\boldsymbol{f}_n(\boldsymbol{a}) &= 0. \end{aligned} \quad (5.6)$$

Interpreting (5.6) as a linear homogeneous system with unknown $\boldsymbol{f}_1(\boldsymbol{a}), \ldots, \boldsymbol{f}_n(\boldsymbol{a})$, not all equal to zero, it follows that the determinant of this system must be zero, showing that the functions $U_1, \ldots, U_n$ cannot be independent in a neighborhood of $\boldsymbol{a}$.

Let us prove that there exist $(n-1)$ independent prime integrals near $\boldsymbol{a} = (a_1, \ldots, a_n)$. Denote by $\boldsymbol{x} = \varphi(\,t;\, \lambda_1, \ldots, \lambda_{n-1}\,)$ the solution of system (5.1) that satisfies the initial condition

$$\boldsymbol{x}(t) = (\lambda_1, \ldots, \lambda_{n-1}, a_n).$$

More explicitly, we have

$$x_i = \varphi_i(t; \lambda_1, \ldots, \lambda_{n-1}), \quad i = 1, 2, \ldots, n. \tag{5.7}$$

We get

$$\begin{aligned} \lambda_i &= \varphi_i(0; \lambda_1, \ldots, \lambda_{n-1}), \quad i = 1, \ldots, n-1, \\ a_n &= \varphi_n(0; \lambda_1, \ldots, \lambda_{n-1}). \end{aligned} \tag{5.8}$$

Using Theorem 3.14, we deduce that the map

$$(t, \lambda_1, \ldots, \lambda_{n-1}) \mapsto \varphi(t, \lambda_1, \ldots, \lambda_{n-1})$$

is a C^1-function of $\lambda_1, \ldots, \lambda_{n-1}$. Moreover, using (5.8), we see that its Jacobian at the point $(0, a_1, \ldots, a_{n-1})$ is

$$\frac{D(\varphi_1, \ldots, \varphi_n)}{D(t, \lambda_1, \ldots, \lambda_{n-1})}(0, a_1, \ldots, a_{n-1}) = \boldsymbol{f}_n(\boldsymbol{a}) \neq 0. \tag{5.9}$$

The inverse function theorem implies that in a neighborhood $\mathcal{V}(\boldsymbol{a})$ of the point $\varphi(0, a_1, \ldots, a_n) = \boldsymbol{a}$ there exist C^1-functions $U_1, \ldots, U_{n-1}, V$ such that

$$\begin{aligned} \lambda_i &= U_i(\boldsymbol{x}), \quad i = 1, \ldots, n-1, \quad \boldsymbol{x} \in \mathcal{V}(\boldsymbol{a}), \\ t &= V(\boldsymbol{x}). \end{aligned} \tag{5.10}$$

By construction, the functions $U_1, \ldots, U_{n-1}, V$ are independent on a neighborhood of $\boldsymbol{a}$ so, in particular, the functions $U_1, \ldots, U_{n-1}$ are independent in a neighborhood of $\boldsymbol{a}$.

Let us prove that $U_1, \ldots, U_{n-1}$ are prime integrals, that is,

$$U_i\big(\varphi(t)\big) = \text{constant}, \quad i = 1, \ldots, n-1,$$

for any solution $\boldsymbol{x} = \varphi(t)$ of system (5.1).

From (5.7), (5.8) and (5.10), it follows that $U_i(\varphi(t)) \equiv \text{constant}$ for all $i = 1, ..., n-1$, and, for any solution $\varphi(t)$ whose initial value is $\varphi(0) = (\tilde{\lambda}, a_n)$, we have $\tilde{\lambda} := (\lambda_1, \ldots, \lambda_{n-1})$. Consider now an arbitrary solution $\boldsymbol{x} = \varphi(t; 0, \boldsymbol{x}_0)$ of the system (5.1) which stays in $\mathcal{V}(\boldsymbol{a})$ and has the value $\boldsymbol{x}_0$ at $t = 0$. As indicated in Sect. 2.5, the uniqueness theorem implies the group property

$$\varphi\big(t + \tau; 0, \boldsymbol{x}_0\big) = \varphi\big(t; 0, \varphi(\tau; 0, \boldsymbol{x}_0)\big). \tag{5.11}$$

On the other hand, for $\boldsymbol{x}_0 \in \mathcal{V}(\boldsymbol{a})$, the system

$$\boldsymbol{x}_0 = \varphi\big(\tau; 0, (\tilde{\lambda}, a_n)\big)$$

has a unique solution $(\tau^0, \tilde{\lambda}^0)$. Using (5.11), we get

$$\varphi(t; 0, \boldsymbol{x}_0) = \varphi\big(t; 0, \varphi(\tau^0; 0, (\tilde{\lambda}^0, a_n))\big) = \varphi(t+\tau^0; 0, (\tilde{\lambda}^0, a_n)) = \psi(t).$$

From the above discussion, we deduce that $U_i(\psi(t)) = \text{constant}$. Hence

$$U_i\big(\varphi(t; 0, \boldsymbol{x}_0)\big) = \text{constant}, \quad i = 1, \dots, n-1.$$

This proves Theorem 5.2. □

Roughly speaking, a system of $(n-1)$-independent prime integrals plays the same role for system (5.1) as a fundamental system of solutions for a linear differential system. This follows from our next theorem.

Theorem 5.3 *Let $U_1, \dots, U_{n-1}$ be prime integrals of system (5.1) which are independent in a neighborhood $\mathcal{V}(\boldsymbol{a})$ of the point $\boldsymbol{a} \in \mathbb{R}^n$. Let W be an arbitrary prime integral of system (5.1) defined on some neighborhood of $\boldsymbol{a}$. Then there exists an open neighborhood $\mathcal{U}$ of the point*

$$\big(U_1(\boldsymbol{a}), \dots, U_{n-1}(\boldsymbol{a})\big) \in \mathbb{R}^{n-1},$$

an open neighborhood $\mathcal{W} \subset \mathcal{V}$ of $\boldsymbol{a}$ and a C^1-function $F : \mathcal{U} \to \mathbb{R}$, such that

$$W(\boldsymbol{x}) = F\big(U_1(\boldsymbol{x}), \dots, U_{n-1}(\boldsymbol{x})\big), \quad \forall \boldsymbol{x} \in \mathcal{W}. \tag{5.12}$$

Proof Fix a function $U_n \in C^1(\mathbb{R}^n)$ such that the system $\{U_1, \dots, U_n\}$ is independent on a neighborhood $\mathcal{V}' \subset \mathcal{V}$ of $\boldsymbol{a}$. The inverse function theorem implies that the C^1-map

$$\mathcal{V}' \ni \boldsymbol{x} \mapsto \Phi(\boldsymbol{x}) := \big(U_1(\boldsymbol{x}), \dots, U_n(\boldsymbol{x})\big) \in \mathbb{R}^n$$

is locally invertible near $\boldsymbol{a}$. This means that there exists an open neighborhood $\mathcal{W}$ of $\boldsymbol{a}$ such that $\Phi\big|_{\mathcal{W}}$ is a bijection onto a neighborhood $\hat{\mathcal{U}}$ of $\Phi(\boldsymbol{a})$ and its inverse $\Phi^{-1} : \hat{\mathcal{U}} \to \mathcal{W}$ is also C^1. The inverse is described by a collection of functions $W_1, \dots, W_n \in C^1(\hat{\mathcal{U}})$,

$$\hat{\mathcal{U}} \ni \boldsymbol{u} := (u_1, \dots, u_n) \mapsto \Phi^{-1}(\boldsymbol{u}) = \big(W_1(\boldsymbol{u}), \dots, W_n(\boldsymbol{u})\big) \in \mathcal{W}.$$

We deduce that

$$x_i = W_i\big(U_1(\boldsymbol{x}), \dots, U_n(\boldsymbol{x})\big), \quad \forall \boldsymbol{x} = (x_1, \dots, x_n) \in \mathcal{W}, \quad i = 1, \dots, n. \tag{5.13}$$

$$u_k = U_k\big(W_1(\boldsymbol{u}), \dots, W_n(\boldsymbol{u})\big), \quad \forall \boldsymbol{u} \in \hat{\mathcal{U}}, \quad k = 1, \dots, n. \tag{5.14}$$

Now, define the function $G \in C^1(\hat{\mathcal{U}})$ by setting

$$G(\boldsymbol{u}) := W\big(W_1(\boldsymbol{u}), \dots, W_n(\boldsymbol{u})\big), \quad \forall \boldsymbol{u} \in \hat{\mathcal{U}}. \tag{5.15}$$

Equalities (5.13) imply that

$$W(\boldsymbol{x}) = G\big(U_1(\boldsymbol{x}), \dots, U_n(\boldsymbol{x})\big), \quad \forall \boldsymbol{x} \in \mathcal{W}. \tag{5.16}$$

On the other hand, $G(\boldsymbol{u}) = G(u_1, \dots, u_n)$ is independent of u_n. Indeed, we have

$$\frac{\partial G}{\partial u_n} = \sum_{i=1}^{n} \frac{\partial W}{\partial W_i} \frac{\partial W_i}{\partial u_n}.$$

By Theorem 5.2, the system $\{W, U_1, \dots, U_{n-1}\}$ is dependent, so *grad* W is a linear combination of *grad* $U_1, \dots,$ *grad* U_{n-1}. Hence, there exist functions $a_1(\boldsymbol{x},), \dots, a_{n-1}(\boldsymbol{x})$, defined on an open neighborhood of $\boldsymbol{a}$, such that

$$\frac{\partial W}{\partial x_i} = \sum_{k=1}^{n-1} a_k(\boldsymbol{x}) \frac{\partial U_k}{\partial x_i}, \quad i = 1, \dots, n.$$

Hence

$$\frac{\partial G}{\partial U_n} = \sum_{i=1}^{n} \left(\sum_{k=1}^{n-1} a_k(\boldsymbol{x}) \frac{\partial U_k}{\partial x_i} \right) \frac{\partial W_i}{\partial u_n} = \sum_{k=1}^{n-1} a_k(\boldsymbol{x}) \left(\sum_{i=1}^{n} \frac{\partial U_k}{\partial x_i} \frac{\partial W_i}{\partial u_n} \right)$$

From (5.14), we deduce that, for any $k = 1, \dots, n-1$,

$$\sum_{i=1}^{n} \frac{\partial U_k}{\partial x_i} \frac{\partial W_i}{\partial u_n} = \frac{\partial u_k}{\partial u_n} = 0.$$

Thus, the function $F(u_1, \dots, u_{n-1}) = G(u_1, \dots, u_{n-1}, u_n)$ satisfies all the postulated conditions. □

5.1.1 Hamiltonian Systems

A mechanical system with n degrees of freedom is completely determined by the vector $\boldsymbol{q}(t) = (q_1(t), \dots, q_n(t))$, representing the *generalized coordinates* of the system, and its derivative $\boldsymbol{q}'(t)$, representing the *generalized velocity*.

The behavior of the system is determined by a function

$$L : \mathbb{R}^{2n} \to \mathbb{R}, \quad L = L(\boldsymbol{q}, \boldsymbol{q}')$$

called the *Lagrangian* of the system. More precisely, according to Hamilton's principle, any trajectory (or motion) of the system during the time interval $[0, T]$ is an extremal of the functional

$$S = \int_0^T L(\boldsymbol{q}(t), \boldsymbol{q}'(t))dt,$$

and, as such, it satisfies the *Euler–Lagrange equation*

$$\frac{d}{dt}\left(\frac{\partial L}{\partial \boldsymbol{q}'}\right) - \frac{\partial L}{\partial \boldsymbol{q}} = 0, \quad \forall t \in [0, T]. \tag{5.17}$$

The functions $p_i := \frac{\partial L}{\partial q_i'}$ are called *generalized momenta*, while the functions $\frac{\partial L}{\partial q_i}$ are called *generalized forces*; see e.g. [13].

The function

$$H : \mathbb{R}^{2n} \to \mathbb{R}, \quad H = H(\boldsymbol{q}, \boldsymbol{p}),$$

defined via the *Legendre transform* (see Appendix A.6)

$$H(\boldsymbol{q}, \boldsymbol{p}) := \sup_{\tilde{\boldsymbol{q}} \in \mathbb{R}^n} \big((\boldsymbol{p}, \tilde{\boldsymbol{q}}) - L(\boldsymbol{q}, \tilde{\boldsymbol{q}}) \big), \tag{5.18}$$

is called the *generalized Hamiltonian*. The definition of H shows that we have the equality

$$H(\boldsymbol{q}, \boldsymbol{p}) + L(\boldsymbol{q}, \tilde{\boldsymbol{q}}) = (\boldsymbol{p}, \tilde{\boldsymbol{q}}) \text{ where } \boldsymbol{p} = \frac{\partial L}{\partial \tilde{\boldsymbol{q}}}(\boldsymbol{q}, \tilde{\boldsymbol{q}}). \tag{5.19}$$

Thus, by setting

$$\boldsymbol{p} := \frac{\partial L}{\partial \boldsymbol{q}'}(\boldsymbol{q}, \boldsymbol{q}')$$

we can rewrite (5.17) as follows

$$\begin{aligned} \boldsymbol{p}'(t) &= -\frac{\partial H}{\partial \boldsymbol{q}}(\boldsymbol{q}(t), \boldsymbol{p}(t)), \\ \boldsymbol{q}'(t) &= \frac{\partial H}{\partial \boldsymbol{p}}(\boldsymbol{q}(t), \boldsymbol{p}(t)), \; t \in [0, T]. \end{aligned} \tag{5.20}$$

System (5.20) is called the *Hamiltonian system* associated with the mechanical system. Theorem 5.1 implies that the Hamiltonian function H is a prime integral of system (5.20). In other words

$$H\big(\boldsymbol{q}(t), \boldsymbol{p}(t)\big) = constant, \tag{5.21}$$

for any trajectory $(\boldsymbol{q}(t), \boldsymbol{p}(t))$ of system (5.20).

In classical mechanics, the Hamiltonian of a system of n-particles of masses $m_1, \ldots, m_n$ has the form

$$H(q_1, \ldots, q_n; p_1, \ldots, p_n) = \sum_{k=1}^{n} \frac{1}{2m_k} p_k^2 + V(q_1, \ldots, q_n), \tag{5.22}$$

where $V(q_1, \ldots, q_n)$ is the potential energy of the system. In other words, $H(\boldsymbol{q}(t), \boldsymbol{p}(t))$ is the total energy of the system at time t, and (5.21) is none other than the conservation of energy law.

The Hamiltonian systems are the most general *conservative* differential systems, that is, for which the energy is a prime integral (see also (1.56) and (1.57)).

In the special case of conservative systems with a single degree of freedom, normalizing the mass to be 1, the Hamiltonian has the form

$$H(q, p) = \frac{1}{2}p^2 + G(q), \quad G(q) = \int_0^q g(r)dr,$$

and system (5.20) reduces to Newton's equation

$$x'' + g(x) = 0, \tag{5.23}$$

or, equivalently, to the system

$$\begin{aligned} x' &= p \\ p' &= -g(x), \end{aligned} \tag{5.24}$$

that we have already investigated. In this case, equality (5.21) becomes

$$\frac{1}{2}x'(t)^2 + G\big(x(t)\big) = C, \tag{5.25}$$

where

$$C := \frac{1}{2}p_0^2 + G(x_0),$$

and (p_0, x_0) are initial data for system (5.24). Integrating (5.25), we get

$$\sqrt{2}\,t = \int_{x_0}^{x(t)} \frac{dr}{\sqrt{C - G(r)}}, \quad x(0) = x_0. \tag{5.26}$$

Equality (5.25) (respectively (5.26)) describes a curve called the *energy level*.

Let us, additionally, assume that g is C^1 and satisfies

$$ug(u) > 0, \quad \forall u \neq 0, \quad g(-u) = -g(u), \quad \forall u \in \mathbb{R}. \tag{5.27}$$

One can prove that, under these assumptions, for C sufficiently small, the solution of (5.26) is periodic; see [3].

Equation (5.23) describes a general class of second-order ODEs. As we have already seen, when g is linear, we obtain the harmonic oscillator equation, while, in the case $g(x) = \sin x$, we obtain the pendulum equation.

If $g(x) = \omega^2 x^2 + \beta x^3$, then (5.23) is called the *Duffing equation*.

5.2 Prime Integrals of Non-autonomous Differential Systems

Consider the differential system

$$\boldsymbol{x}' = \boldsymbol{f}(t, \boldsymbol{x}), \tag{5.28}$$

where $\boldsymbol{f} : \Omega \subset \mathbb{R}^{n+1} \to \mathbb{R}^n$ is continuous, differentiable with respect to $\boldsymbol{x}$, and the derivative $\boldsymbol{f}_{\boldsymbol{x}}$ is continuous on the open set Ω. Imitating the preceding section, we will say that the function $V = V(t, \boldsymbol{x}) : \Omega \to \mathbb{R}$ is a *prime integral* of the system (5.28) on an open subset $\Omega_0 \subset \Omega$ if V is C^1 on Ω_0, it is not identically constant on Ω_0 and $V(t, \varphi(t)) = constant$ for any solution $\varphi(t)$ of (5.28) whose graph is contained in Ω_0. The proof of the following characterization theorem is similar to the proof of Theorem 5.1 and, therefore, we omit it.

Theorem 5.4 *The C^1-function on Ω_0 is the prime integral of system (5.28) if and only if it satisfies the equality*

$$\frac{\partial V}{\partial t}(t, \boldsymbol{x}) + \sum_{i=1}^{n} \frac{\partial V}{\partial x_i}(t, \boldsymbol{x}) f_i(t, \boldsymbol{x}) = 0, \quad \forall (t, \boldsymbol{x}) \in \Omega_0. \tag{5.29}$$

To prove other properties of the prime integrals of system (5.28), it suffices to observe that this system can be regarded as an $(n + 1)$-dimensional *autonomous* differential system.

Indeed, interpreting $(t, \boldsymbol{x})$ as unknown functions and introducing a new real variable s, we can rewrite (5.28) in the form

$$\begin{aligned} \frac{dx}{ds} &= \boldsymbol{f}(t, \boldsymbol{x}), \\ \frac{dt}{ds} &= 1. \end{aligned} \tag{5.30}$$

In this fashion, a prime integral of (5.28) becomes a prime integral of the autonomous system (5.30). Theorems 5.2 and 5.3 imply the following result.

Theorem 5.5 *In a neighborhood of the point* $(t_0, \boldsymbol{a}_0) \in \Omega$, *system (5.28) admits exactly n independent prime integrals* $V_1, \dots, V_n$. *Any other prime integral* $V(t, \boldsymbol{x})$ *has the form*

$$V(t, \boldsymbol{x}) = F\big(V_1(t, \boldsymbol{x}), \dots, V_n(t, \boldsymbol{x})\big), \tag{5.31}$$

where $F(v_1, \dots, v_n)$ *is a differentiable function defined in a neighborhood of the point* $\big(V_1(t_0, \boldsymbol{a}_0), \dots, V_n(t_0, \boldsymbol{a}_0)\big) \in \mathbb{R}^n$.

The knowledge of k independent prime integrals $k < n$ allows the reduction of the dimension of the system. Indeed, if $U_1(t, \boldsymbol{x}), \dots, U_k(t, \boldsymbol{x})$ are k independent prime integrals of system (5.28), then, locally, we have the equalities

$$\begin{aligned} U_1(t, \boldsymbol{x}_1, \dots, x_n) &= C_1 \\ U_2(t, \boldsymbol{x}_1, \dots, x_n) &= C_2 \\ \vdots \qquad & \;\; \vdots \; \vdots \\ U_k(t, \boldsymbol{x}_1, \dots, x_n) &= C_k, \end{aligned} \tag{5.32}$$

where $\boldsymbol{x}(t) = (x_1(t), \dots, x_n(t))$ is a trajectory of the system, and $C_1, \dots, C_k$ are constants.

Since the functions $U_1, \dots, U_k$ are independent, we may assume that the functional determinant

$$\frac{D(U_1, \dots, U_k)}{D(x_1, \dots, x_k)}$$

is nonzero. In other words, the implicit system (2.43) can be solved with respect to $(x_1, \dots, x_k)$; see Theorem A.3. We deduce that

$$\begin{aligned} x_1 &= \varphi_1(t, x_{k+1}, \dots, x_n; C_1, \dots, C_k) \\ x_2 &= \varphi_2(t, x_{k+1}, \dots, x_n; C_1, \dots, C_k) \\ \vdots \; &\vdots \qquad\qquad \vdots \\ x_k &= \varphi_k(t, x_{k+1}, \dots, x_n; C_1, \dots, C_k). \end{aligned} \tag{5.33}$$

In this fashion, the only unknown variables left in (2.39) are $x_{k+1}, \dots, x_n$. In particular, the knowledge of n independent prime integrals of the system is equivalent to solving it.

Example 5.1 Consider, for example, the differential system,

$$\begin{aligned} x_1' &= x_2^2 \\ x_2' &= x_1 x_2. \end{aligned}$$

Rewriting it in a symmetric form

$$\frac{dx_1}{x_2^2} = \frac{dx_2}{x_1 x_2},$$

we observe that $U(x_1, x_2) = x_1^2 - x_2^2$ is a prime integral. In other words, the general solution admits the representation

$$x_1^2 - x_2^2 = C, \tag{5.34}$$

where C is a real constant. An explicit form of the solution in the space (x_1, x_2, t) can be found by setting $x_2 = \sqrt{x_1'}$ and using this in (5.34). We obtain in this fashion a differential equation in x_1 and t which, upon solving, yields an explicit description of x_1 as a function of t.

5.3 First-Order Quasilinear Partial Differential Equations

In this section, we will investigate the equation

$$\sum_{i=1}^{n} a_i(\boldsymbol{x}, z) z_{x_i} = a(\boldsymbol{x}, z), \quad \boldsymbol{x} = (x_1, \dots, x_n), \tag{5.35}$$

with the unknown function $z = z(\boldsymbol{x})$, where for $i = 1, \dots, n$ the functions a_i are C^1 on an open set $\Omega \subset \mathbb{R}^{n+1}$, and satisfy the condition

$$\sum_{i=1}^{n} a_i(\boldsymbol{x}, z)^2 \neq 0, \quad \forall (\boldsymbol{x}, z) \in \Omega. \tag{5.36}$$

We denote by z_{x_i} the partial derivatives $\frac{\partial z}{\partial x_i}$, $i = 1, \dots, n$. Equation (5.35) is called a *first-order, quasilinear partial differential equation.*

Definition 5.2 A solution of (5.35) on an open set $D \subset \mathbb{R}^n$ is a function $z \in C^1(D)$ that satisfies equality (5.35) for all $\boldsymbol{x} \in D$.

Geometrically, the graph of a solution of (5.35) is a hypersurface in $\mathbb{R}^{n+1}$ with the property that the vector field $(a_1, \dots, a_n, a)$ is tangent to this hypersurface at all of its points.

We associate with (5.35) the system of ODEs

$$\frac{dz}{da} = a(\boldsymbol{x}, z), \quad \frac{dx_i}{ds} = a_i(\boldsymbol{x}, z), \quad i = 1, \dots, n, \tag{5.37}$$

called the *characteristics equation*. The solutions of (5.37), for which the existence theory presented in Sect. 2.1 applies, are called *the characteristic curves* of Eq. (5.35).

We seek a solution of (5.35) described implicitly by an equation of the form

$$u(\boldsymbol{x}, z) = 0.$$

Then

$$z_{x_i} = -\frac{u_{x_i}}{u_z}, \quad \forall i = 1, \dots, n,$$

and thus we can rewrite (5.35) in the form

$$\sum_{i=1}^{n} a_i(\boldsymbol{x}, z)u_{x_i} + a(\boldsymbol{x}, z)u_z = 0, \quad (\boldsymbol{x}, z) \in \Omega. \tag{5.38}$$

Theorem 5.1 characterizing the prime integrals of autonomous systems of ODEs shows that a function u is a solution of (5.38) if and only if it is a prime integral of the characteristics equation (5.37).

Given our assumptions on the functions a_i and a, we deduce that system (5.37) admits n independent prime integrals $U_1, \dots, U_n$ on an open subset $\Omega' \subset \Omega$. The general solution of (5.38) has the form (see Theorem 5.5)

$$u(\boldsymbol{x}, z) = F\big(U_1(\boldsymbol{x}, z), \dots, U_n(\boldsymbol{x}, z)\big), \quad (\boldsymbol{x}, z) \in \Omega', \tag{5.39}$$

where F is an arbitrary C^1-function. Thus, solving (5.38), and indirectly (5.35), reduces to solving the characteristics equation, that is, finding n independent prime integrals of system (5.37).

Example 5.2 Consider the following first-order quasilinear PDE

$$x_1 z z_{x_1} + x_2 z_{x_2} = -x_1 x_2. \tag{5.40}$$

The characteristics equation is given by the system

$$\frac{dx_1}{ds} = x_1 z, \quad \frac{dx_2}{ds} = x_2 z, \quad \frac{dz}{ds} = -x_1 x_2,$$

or, in symmetric form,

$$\frac{dx_1}{x_1 z} = \frac{dx_2}{x_2 z} = \frac{dz}{x_1 x_2}. \tag{5.41}$$

From the first equality, we deduce that

$$\frac{dx_1}{x_1} = \frac{dx_2}{x_2}.$$

Hence

$$U_1(x_1, x_2, z) = \frac{x_1}{x_2}$$

is a prime integral of the characteristics system. System (5.41) also implies the equality

$$2z dz = -d(x_1 x_2).$$

Hence, the function

$$U_2(x_1, x_2, z) = z^2 + x_1x_2$$

is another prime integral of (5.41) that is obviously independent of U_1. Thus, the general solution of (5.40) is given implicitly by the equation

$$F\left(\frac{x_1}{x_2}, z^2 + x_1x_2\right) = 0, \quad (x_1, x_2) \in \mathbb{R}^2,$$

where $F : \mathbb{R}^2 \to \mathbb{R}$ is an arbitrary C^1-function.

As in the case of systems of ODEs, when investigating partial differential equations, we are especially interested in solutions satisfying additional conditions or having prescribed values on certain parts of their domains of definition. In the remainder of this section, we will study one such condition, which is a natural generalization of the initial condition in the case of ODEs.

5.3.1 The Cauchy Problem

In the space $\mathbb{R}^{n+1}$, consider the $(n-1)$-dimensional submanifold Γ defined by the equations

$$\begin{aligned} x_i &= \varphi_i(u_1, \dots, u_{n-1}), \quad i = 1, \dots, n, \\ z &= \varphi(u_1, \dots, u_{n-1}), \quad \boldsymbol{u} := (u_1, \dots, u_{n-1}) \in U \subset \mathbb{R}^{n-1}, \end{aligned} \tag{5.42}$$

where φ and φ_i are C^1-functions defined on the open set U. We will assume that

$$\det\left(\frac{\partial \varphi_i}{\partial u_j}\right)_{1 \le i,j \le n-1} \neq 0.$$

A solution of Eq. (5.35) satisfying the *Cauchy condition* (5.42) is a solution of (5.35) whose graph contains the manifold Γ, that is,

$$\varphi(\boldsymbol{u}) = z\big(\varphi_1(\boldsymbol{u}), \dots, \varphi_n(\boldsymbol{u})\big), \quad \forall \boldsymbol{u} \in U. \tag{5.43}$$

We will prove the following existence result.

Theorem 5.6 *Suppose that the following nondegeneracy condition is satisfied*

$$\Delta = \det \begin{pmatrix} a_1(\varphi_1(\boldsymbol{u}), ..., \varphi_n(\boldsymbol{u}), \varphi(\boldsymbol{u})) & \cdots & a_n(\varphi_1(\boldsymbol{u}), ..., \varphi_n(\boldsymbol{u}), \varphi(\boldsymbol{u})) \\ \dfrac{\partial \varphi_1}{\partial u_1}(\boldsymbol{u}) & \cdots & \dfrac{\partial \varphi_n}{\partial u_1}(\boldsymbol{u}) \\ \vdots & \vdots & \vdots \\ \dfrac{\partial \varphi_{n-1}}{\partial u_1}(\boldsymbol{u}) & \cdots & \dfrac{\partial \varphi_n}{\partial u_{n-1}}(\boldsymbol{u}) \end{pmatrix} \neq 0, \quad (5.44)$$

$\forall \boldsymbol{u} \in U$. Then, the Cauchy problem (5.35), (5.42) has a unique solution defined in a neighborhood of the manifold Γ.

Proof Let us observe that the characteristics equation (5.37) define locally an $(n+1)$-dimensional family of solutions

$$x_i = x_i(s; s_0, \boldsymbol{x}_0, z_0), \quad i = 1, \dots, n,$$
$$z = z(s; s_0, \boldsymbol{x}_0, z_0), \quad s \in I,$$

where $I \subset \mathbb{R}$ is an interval. In the above equations, we let $(\boldsymbol{x}_0, z_0) \in \Gamma$, that is,

$$\boldsymbol{x}_0 = \big(\varphi_1(\boldsymbol{u}), \dots, \varphi_n(\boldsymbol{u})\big), \quad z_0 = \varphi(\boldsymbol{u}), \quad \boldsymbol{u} \in U.$$

The quantities $(s, \boldsymbol{u})$ are solutions of the nonlinear system

$$\begin{aligned} x_i &= x_i\big(s; s_0, \varphi_1(\boldsymbol{u}), \dots, \varphi_n(\boldsymbol{u})\big), \\ z &= z\big(s; s_0, \varphi_1(\boldsymbol{u}), \dots, \varphi_n(\boldsymbol{u})\big), \quad \boldsymbol{u} \in U. \end{aligned} \tag{5.45}$$

From the characteristics equation (5.37), we deduce that

$$\frac{D(x_1, \dots, x_n)}{D(s, u_1, \dots, u_{n-1})} = \Delta \neq 0 \ \text{ for } \ s = s_0, \quad \boldsymbol{x} = \boldsymbol{x}_0.$$

The inverse function theorem shows that the correspondence

$$(s, u_1, \dots, u_n) \mapsto (x_1, \dots, x_n)$$

defined by (5.45) is a diffeomorphism of an open set in the $(s, \boldsymbol{u})$-space onto a neighborhood of Γ. Thus, there exist C^1-functions $\Phi, \Phi_1, \dots, \Phi_{n-1}$ defined on a neighborhood of Γ such that

$$\begin{aligned} u_i &= \Phi_i(x_1, \dots, x_n), \quad i = 1, \dots, n-1, \\ s &= \Phi(x_1, \dots, x_n). \end{aligned} \tag{5.46}$$

This expresses the quantities $u_1, \dots, u_n$ as functions of $\boldsymbol{x}$ and, using this in the second equation in (5.45), we obtain a function $z = z(\boldsymbol{x})$ whose graph, by design, contains Γ. We want to prove that z is a solution of (5.35).

Indeed, by construction, z is part of the solution of the characteristics equation (5.37) so that

$$\frac{dz}{ds} = a(\boldsymbol{x}, z).$$

On the other hand, from equalities (5.37) and (5.45), we deduce that

$$\frac{dz}{ds} = \sum_{i=1}^{n} \frac{\partial z}{\partial z_i} \frac{dx_i}{ds} = \sum_{i=1}^{n} z_{x_i} a_i(\boldsymbol{x}, z).$$

This shows that z is a solution of (5.35).

The uniqueness follows from the fact that any solution of the Cauchy problem (5.35), (5.42) is necessarily obtained by the above procedure. Indeed, if $\tilde{z}(x_1, \dots, x_n)$ is another solution of the Cauchy problem (5.35), (5.42), then, for any initial vector $(s_0, \boldsymbol{x}_0) \in \mathbb{R}^{n+1}$, the Cauchy problem

$$\begin{aligned} \frac{dx_i}{ds} &= a_i\big(\boldsymbol{x}, \tilde{z}(\boldsymbol{x})\big), \quad && i = 1, \dots, n, \\ x_i(s_0) &= x_i^0, && i = 1, \dots, n, \end{aligned} \tag{5.47}$$

admits a local solution $\boldsymbol{x} = \boldsymbol{x}(s)$. Since $\tilde{z}(\boldsymbol{x})$ is a solution of (5.35), we deduce that

$$\frac{d}{ds}\tilde{z}(\boldsymbol{x}(s)) = \sum_{i=1}^{n} \tilde{z}_{x_i}(\boldsymbol{x}(s)) = a(\boldsymbol{x}, \tilde{z}(\boldsymbol{x})).$$

In other words, the curve

$$s \mapsto \big(\boldsymbol{x}(s), z(\boldsymbol{x}(s))\big)$$

is a characteristic curve of Eq. (5.35). If $x_i^0 = \varphi_i(\boldsymbol{u})$, $i = 1, \dots, n$, then necessarily $\tilde{z}(\boldsymbol{x}(s_0)) = \varphi(\boldsymbol{u})$. From the uniqueness of the Cauchy problem for system (5.37), we see that

$$\tilde{z}\big(x_1(s), \dots, x_n(s)\big) = z\big(x_1(s), \dots, x_n(s)\big),$$

where $z = z(\boldsymbol{x})$ is the solution constructed earlier via equations (5.45). Hence $z = \tilde{z}$. □

Remark 5.1 Hypothesis (5.44) is essential for the existence and uniqueness of the Cauchy problem for (5.35). If $\Delta = 0$ along the submanifold Γ, then the Cauchy problem admits a solution only if Γ is a *characteristic submanifold* of Eq. (5.35), that is, at every point of Γ the vector field

$$\big(a_1(\boldsymbol{x}, z), \dots, a_n(\boldsymbol{x}, z), a(\boldsymbol{x}, z)\big)$$

is tangent to Γ. However, in this case the solution is not unique.

Let us also mention that, in the case $n = 2$, Eq. (5.35) reduces to

$$P(x, y, z)z_x + Q(x, y, z)z_y = R(x, y, z), \quad (5.48)$$

and the Cauchy problem consists in finding a function $z = z(x, y)$ whose graph contains the curve

$$x = \varphi(u), \quad y = \psi(u), \quad z = \chi(u), \quad u \in I,$$

where P, Q, R are C^1-functions on a domain of $\mathbb{R}^3$, and φ, ψ, χ are C^1-functions on an interval $I \subset \mathbb{R}$.

Example 5.3 Let us find a function $z = z(x, y)$ satisfying the first-order quasilinear PDE

$$xz_x + zz_y = y,$$

and such that its graph contains the line

$$y = 2z, \quad x + 2y = -z. \quad (5.49)$$

This line can be parameterized by the equations

$$x = -3u, \quad y = 2u, z = u, \quad u \in I,$$

and we observe that assumption (5.44) in Theorem 5.6 is satisfied for $u \neq 0$. The characteristics equation is

$$\frac{dx}{ds} = x, \quad \frac{dy}{ds} = z, \quad \frac{dz}{ds} = y,$$

and its general solution is $x = x_0 e^s$, $y = \frac{1}{2}\left(e^s(y_0 + z_0) + e^{-s}(y_0 - z_0)\right)$, $z = \frac{1}{2}$ $\left(e^s(y_0 + z_0) - e^{-s}(y_0 - z_0)\right)$. The graph of z is filled by the characteristic curves originating at points on the line (5.49) and, as such, it admits the parametrization

$$x = -3ue^s, \quad y = \frac{u}{2}\left(3e^s + e^{-s}\right), \quad z = \frac{u}{2}\left(3e^s - e^{-s}\right).$$

5.4 Conservation Laws

A large number of problems in physics lead to partial differential equations of the form

$$z_x + a(z)z_y = 0, \quad (x, y) \in [0, \infty) \times \mathbb{R}, \quad (5.50)$$

with the Cauchy condition

$$z(0, y) = \varphi(y), \quad \forall y \in \mathbb{R}. \tag{5.51}$$

Here, $a, \varphi : \mathbb{R} \to \mathbb{R}$ are C^1-functions. Equation (5.50) is known as a *conservation law equation* and arises most frequently in the mathematical modeling of certain dynamical phenomena that imply the conservation of certain quantities such as mass, energy, momentum, etc.

If we denote by $\rho(y, t)$ the density of that quantity at the point $y \in \mathbb{R}$ and at the moment of time $t \geq 0$, and by $q(y, t)$ the flux per unit of time, then the conservation law for that quantity takes the form

$$\frac{d}{dt}\int_{y_1}^{y_2} \rho(y, t)dy + q(y_2, t) - q(y_1, t) = 0.$$

Letting $y_2 \to y_1$, we deduce that

$$\frac{\partial \rho}{\partial t} + \frac{\partial q}{\partial y} = 0, \quad y \in \mathbb{R}, \quad t \geq 0. \tag{5.52}$$

If the flux q is a function of ρ, that is,

$$q = Q(\rho), \tag{5.53}$$

then equation (5.52) becomes an equation of type (5.50)

$$\rho_t + Q'(\rho)\rho_y = 0. \tag{5.54}$$

Let us illustrate this abstract model with several concrete examples (see, e.g., [19]).

Example 5.4 (*Large waves in rivers*) Consider a rectangular channel directed along the y-axis and of constant width. Denote by $q(y, t)$ the flux of water per unit of width and by $h(y, t)$ the height of the water wave in the channel at the point y and moment t. The conservation of mass leads to (5.52) where $q = h$. Between the flux and the depth h, we have a dependency of type (5.53)

$$q = Q(h),$$

and, experimentally, it is found that Q is given by

$$Q(h) = \alpha h^{\frac{3}{2}}, \quad h \geq 0.$$

In this case, the conservation of mass equation becomes

$$h_t + \frac{3\alpha}{2} h^{\frac{1}{2}} h_y = 0. \tag{5.55}$$

The same Eq. (5.54) models the movement of icebergs. In this case, Q has the form

$$Q(h) = Ch^N, \quad N \in (3, 5).$$

Example 5.5 (*Traffic flow*) Consider the traffic flow of cars on a highway directed along the y-axis. If $\rho(y, t)$ is the density of cars (number of cars per unit of length) and v is the velocity, then the flux is given by $q(y, t) = \rho(y, t)v(y, t)$. If the velocity is a function of ρ, $v = V(\rho)$, then the conservation law equation leads to

$$\rho_t + W(\rho)\rho_y = 0,$$

where $W(\rho) = V(\rho) + \rho V'(\rho)$.

Let us now return to the Cauchy problem (5.50), (5.51) and try to solve it using the method of characteristics described in the previous section. The characteristics equation for (5.50) has the form

$$\frac{dx}{ds} = 1, \quad \frac{dy}{ds} = a(z), \quad \frac{dz}{ds} = 0, \tag{5.56}$$

while the curve that appears in (5.51) has the parametric description

$$x = 0, \quad y = t, \quad z = \varphi(t). \tag{5.57}$$

The general solution of (5.56) is

$$x = s + x_0, \quad y = a(z_0)s + y_0, \quad z = z_0.$$

Thus, the graph of z admits the parametrization,

$$x = s, \quad y = a\big(\varphi(t)\big)s + t, \quad z = \varphi(t),$$

or, equivalently,

$$z = \varphi\big(y - xa(z)\big), \quad x \geq 0, \quad y \in \mathbb{R}. \tag{5.58}$$

According to the implicit function theorem (Theorem A.3), Eq. (5.58) defines a C^1-function z in a neighborhood of any point (x_0, y_0), such that

$$x_0\varphi'\big(y_0 - x_0a(z_0)\big)a'(z_0) \neq 1.$$

Thus, there exists a unique solution $z = z(x, y)$ to (5.50)–(5.51) defined on a tiny rectangle $[0, \delta] \times [-b, b] \subset \mathbb{R}^2$.

From the above construction, we can draw several important conclusions concerning the solution of (5.50). We think of the variable x as time and, for this reason, we will relabel it t. Observe first that, if the function $\varphi : \mathbb{R} \to \mathbb{R}$ is bounded,

$$\big|\,\varphi(y)\,\big| \le M, \quad \forall y \in \mathbb{R},$$

then (5.58) shows that the value of z at the point $(t, \bar{y})$ depends only on the initial condition on the interval

$$\big\{w \in \mathbb{R};\quad |w - \bar{y}| \le Ct\big\},$$

where

$$C := \sup_{|z| \le M} |a(z)|.$$

In particular, if the initial data φ is supported on the interval $[-R, R]$, then the solution $z(t, y)$ is supported in the region

$$\big\{(t, y);\quad |y| \le R + Ct\big\},$$

that is, $z(t, y) = 0$ for $|y| > R + Ct$. This property of the solution is called *finite speed propagation*.

An interesting phenomenon involving the solutions of the conservation law equations is the appearance of singularities. More precisely, for large values of t, the function $z = z(t, y)$ defined by (5.58), that is,

$$z = \varphi(y - ta(z)),\ t \ge 0,\ y \in \mathbb{R}, \tag{5.59}$$

can become singular and even multivalued.

Take, for example, Eq. (5.55) where $\varphi(y) = y^2 + 1$. In this case, the equation (5.59) becomes

$$z = \left(y - \frac{3\alpha}{2} t z^{\frac{1}{2}}\right)^2 + 1.$$

The above equation describes a surface in the (t, y, z)-space which is the graph of a function $z = z(t, y)$ provided that

$$0 < t < \frac{2\sqrt{y^2 + 1}}{3\alpha y},\quad y > 0.$$

The solution becomes singular along the curve

$$3\alpha t y = 2\sqrt{y^2 + 1}.$$

It is interesting to point out that the formation of this singularity is in perfect agreement with physical reality. Let us recall that, in this case, $z = h(t, y)$ represents the height of the water wave at time t and at the point y in the channel, and $\varphi(y)$ describes the initial shape of the wave; see the left-hand side of Fig. 5.1.

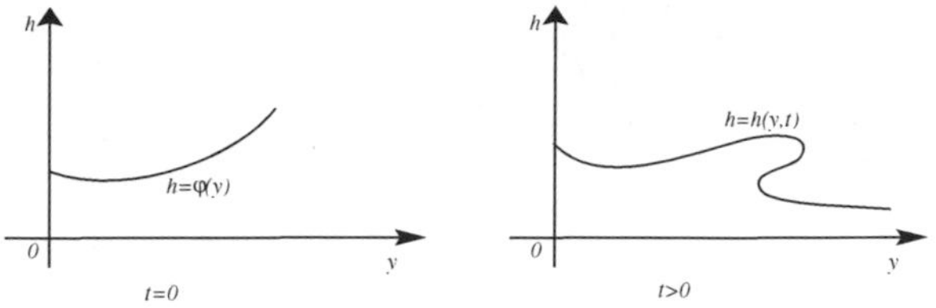

Fig. 5.1 The evolution of water waves

At the location y along the channel, the wave will break at time

$$T(y) = \frac{2\sqrt{y^2+1}}{3\alpha y}.$$

This corresponds to the formation of a singularity in the function $y \mapsto h(t, y)$; see the right-hand side of Fig. 5.1.

This example shows that the formation of singularities in the solutions to (5.50) is an unavoidable reality that has physical significance. Any theory of these types of equations would have to take into account the existence of solutions that satisfy the equations in certain regions and become singular along some curves in the (t, y)-plane. Such functions, which are sometimes called "shocks", cannot be solutions in the classical sense and force upon us the need to extend the concept of solutions for (5.50) using the concept of distribution or generalized function, as we did in Sect. 3.8.

Definition 5.3 A locally integrable function in the domain

$$D = \big\{\, (x, y) \in \mathbb{R}^2;\ \ x \geq 0 \,\big\}$$

is called a *weak solution* of equation (5.50) if, for any C^1-function ψ with compact support on D, we have

$$\int_D \big(z\psi_x + \psi A(z) \big) dxdy + \int_{-\infty}^{\infty} \varphi(y)\psi(0, y)dy = 0, \tag{5.60}$$

where

$$A(z) := \int_0^z a(r)dr.$$

Integrating by parts, we see that any C^1-solution of (5.50) (let's call it a *classical solution*) is also a weak solution. On the other hand, a weak solution need not even be continuous. Let us remark that, when dealing with weak solutions, we can allow the initial function $\varphi(y)$ to have discontinuities. Such situations can appear frequently in real life examples.

Let us assume that the weak solution z of (5.50) is C^1 outside a curve

$$\Gamma = \big\{\, (x, y) \in D;\ \ y = \ell(x) \,\big\}.$$

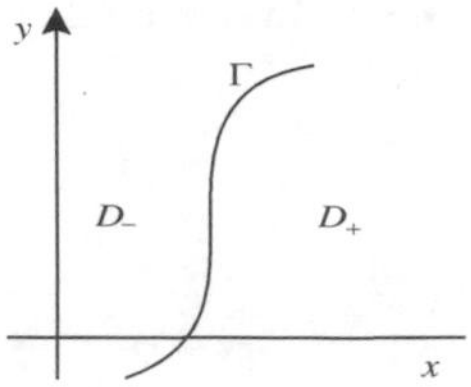

Fig. 5.2 A weak solution with singularities along a curve Γ

We denote by D_- and D_+ the two regions in which D is divided by Γ; see Fig. 5.2.

If, in (5.60), we first choose ψ to have support in D_- and then to have support in D_+, we deduce that

$$z_x(x,y) + a(z) + z_y(x,y) = 0, \quad \forall (x,y) \in D_-, \tag{5.61}$$

$$z_x(x,y) + a(z) + z_y(x,y) = 0, \quad \forall (x,y) \in D_+, \tag{5.62}$$

$$z(0,y) = \varphi(y), \quad \forall y \in \mathbb{R}. \tag{5.63}$$

We set

$$z^{\pm}(x,\ell(x)) = \lim_{\substack{(x_1,y_1)\to(x,\ell(x))\\ (x_1,y_1)\in D_\pm}} z(x_1,\ell(x_1)).$$

Let ψ be an arbitrary C^1 function with compact support on D. Multiplying successively each of equations (5.61) and (5.62) by ψ, and integrating on D_- and respectively D_+, we deduce that

$$\begin{aligned}
0 &= \int_{D_-} \big(z_x + A(z)z_y\big)\psi dxdy = -\int_{D_-} \big(z\psi_x + A(z)\psi_y\big)dxdy \\
&\quad + \int_0^\infty \big(\ell'(x)z^- - A(z^-)\big)\psi(x,\ell(x))dx - \int_0^\infty \varphi(y)\psi(0,y)dy. \\
0 &= \int_{D_+} \big(z_x + A(z)z_y\big)\psi dxdy = -\int_{D_-} \big(z\psi_x + A(z)\psi_y\big)dxdy \\
&\quad - \int_0^\infty \big(\ell'(x)z^+ - A(z^+)\big)\psi(x,\ell(x))dx.
\end{aligned}$$

If we add the last two equalities and then use (5.60), we deduce that

$$\int_0^\infty \Big(\big(\ell'(x)z^- - A(z^-)\big) - \big(\ell'(x)z^+ - A(z^+)\big)\Big)\psi(x,\ell(x))dx = 0.$$

Since ψ is arbitrary, we deduce the pointwise equality

$$A(z^+(x,\ell(x))) - A(z^-(x,\ell(x))) = \ell'(x)(z^+(x,\ell(x)) - z^+(x,\ell(x))), \tag{5.64}$$

$\forall x \geq 0$. In other words, along Γ we have the *jump* condition

$$A(z^+) - A(z^-) = \nu(z^+ - z^-), \tag{5.65}$$

where ν is the slope of the curve. Equality (5.65) is called the *Rankine–Hugoniot relation* and describes the jump in velocity when crossing a shock curve; see [7].

Example 5.6 Consider the equation

$$z_x + z^2 z_y = 0 \tag{5.66}$$

satisfying the Cauchy condition

$$z(0, y) = \begin{cases} 0, & y \leq 0, \\ 1, & y > 0. \end{cases} \tag{5.67}$$

Let us observe that the function

$$z(x, y) := \begin{cases} 0, & y \leq \frac{x}{3}, \\ 1, & y > \frac{x}{3}, \end{cases} \tag{5.68}$$

is a weak solution for the Cauchy problem (5.66), (5.67). Indeed, in this case, equality (5.60)

$$\int_0^\infty dx \int_{\frac{x}{3}}^\infty \left(\psi_x(x, y) + \frac{1}{3}\psi_y(x, y) \right) dy + \int_0^\infty \psi(0, y) dy = 0$$

is obviously verified.

The weak solutions are not unique in general. Besides (5.68), the Cauchy problem (5.66), (5.67) also admits the solution

$$z(x, y) := \begin{cases} 0, & \frac{y}{x} < 0, \\ \left(\frac{y}{x}\right)^{\frac{1}{2}}, & 0 \leq \frac{y}{x} \leq 1, \\ 1, & \frac{y}{x} > 1. \end{cases}$$

The nonuniqueness of weak solutions requires finding criteria that will select from the collection of all possible weak solutions for a given problem those that have a physical significance. A criterion frequently used is the *entropy criterion* according to which we choose only the solutions for which the entropy of the system is increasing.

5.5 Nonlinear Partial Differential Equations

In this section, we will investigate first-order nonlinear PDEs of the form

$$F(x_1, x_2, \dots, x_n, z, p_1, p_2, \dots, p_n) = 0, \tag{5.69}$$

where F is a C^2 function on an open set $\Omega \subset \mathbb{R}^{2n+1}$ and we denote by p_i the functions

$$p_i(x_1, \dots, x_n) := z_{x_i}(x_1, \dots, x_n), \quad i = 1, 2, \dots, n.$$

(We note that (5.35) is a particular case of (5.69).)

By a solution of (5.69), we understand a C^1-function $z = z(x_1, \dots, x_n)$ defined on an open set $D \subset \mathbb{R}^n$ and satisfying (5.69) for all $\boldsymbol{x} = (x_1, \dots, x_n) \in D$. Such a solution is called an *integral manifold* of equation (5.69).

Consider an $(n-1)$-dimensional submanifold Γ of the Euclidean space $\mathbb{R}^{n+1}$ with coordinates $(x_1, \dots, x_n, z)$ described parametrically by equations (5.42), that is,

$$\begin{aligned} x_i &= \varphi_i(\boldsymbol{u}), \quad i = 1, \dots, n, \quad \boldsymbol{u} = (u_1, \dots, u_{n-1}) \in U, \\ z &= \varphi(\boldsymbol{u}). \end{aligned} \tag{5.70}$$

As in the case of quasilinear partial differential equations, we define a solution of the Cauchy problem associated with equation (5.69) and the manifold Γ to be a solution of (5.69) whose graph contains the manifold Γ. In the sequel, we will use the notations

$$Z := F_z, \quad X_i = F_{x_i}, \quad P_i := F_{p_i}, \quad i = 1, \dots, n,$$

and we will impose the nondegeneracy condition

$$\det \begin{bmatrix} P_1 & P_2 & \cdots & P_n \\ \dfrac{\partial \varphi_1}{\partial u_1} & \dfrac{\partial \varphi_2}{\partial u_1} & \cdots & \dfrac{\partial \varphi_n}{\partial u_1} \\ \vdots & \vdots & \vdots & \vdots \\ \dfrac{\partial \varphi_1}{\partial u_{n-1}} & \dfrac{\partial \varphi_2}{\partial u_{n-1}} & \cdots & \dfrac{\partial \varphi_n}{\partial u_{n-1}} \end{bmatrix} \neq 0 \ \text{ on } U. \tag{5.71}$$

Theorem 5.7 *Under the above assumptions, the Cauchy problem (5.69), (5.70) has a unique solution defined in a neighborhood of the manifold Γ.*

Proof The proof has a constructive character and we will highlight a method known in literature as the *method of characteristics* or *Cauchy's method*, already used for equation (5.35).

We associate to equation (5.69) a system of ODEs, the so-called characteristics equation

$$\begin{aligned} \frac{dx_i}{ds} &= P_i, \quad i = 1, \dots, n, \\ \frac{dz}{ds} &= \sum_{i=1}^{n} p_i P_i, \quad s \in I, \\ \frac{dp_i}{ds} &= -(p_i Z + X_i), \quad i = 1, \dots, n. \end{aligned} \tag{5.72}$$

Since the function P_i, X_i and Z are C^1, thc existence and uniqueness theorem implies that, for any $s_0 \in I$ and $(\boldsymbol{x}^0, z^0, \boldsymbol{p}^0) \in \mathbb{R}^{2n+1}$, system (5.72) with the initial conditions

$$x_i(s_0) = x_i^0, \quad z(s_0) = z^0, \quad p_i(s_0) = p_i^0,$$

admits a unique solution

$$\begin{aligned} x_i &= x_i(s; s_0, \boldsymbol{x}^0, z^0, \boldsymbol{p}^0), \ i = 1, \dots, n \\ z &= z(s; s_0, \boldsymbol{x}^0, z^0, \boldsymbol{p}^0), \\ p_i &= p_i(s; s_0, \boldsymbol{x}^0, z^0, \boldsymbol{p}^0), \ i = 1, \dots, n. \end{aligned} \tag{5.73}$$

The function $s \mapsto (x_i(s), z(s))$ is called a characteristic curve.

The integral manifold we seek is determined by the family of characteristic curves originating at the moment $s = s^0$ at points on the initial manifold Γ (see Fig. 5.3). Thus, we are led to define

$$\begin{aligned} x_i^0 &= \varphi_i(u_1, \dots, u_{n-1}), \quad i = 1, \dots, n, \\ z^0 &= \varphi(u_1, \dots, u_{n-1}). \end{aligned} \tag{5.74}$$

The fact that the function whose graph is this surface has to satisfy (5.69) adds another constraint,

$$F\big(\varphi_{(\boldsymbol{u})}, \dots, \varphi_n(\boldsymbol{u}), \varphi(\boldsymbol{u}), p_1^0, \dots, p_n^0\big) = 0, \quad \forall \boldsymbol{u} \in U. \tag{5.75}$$

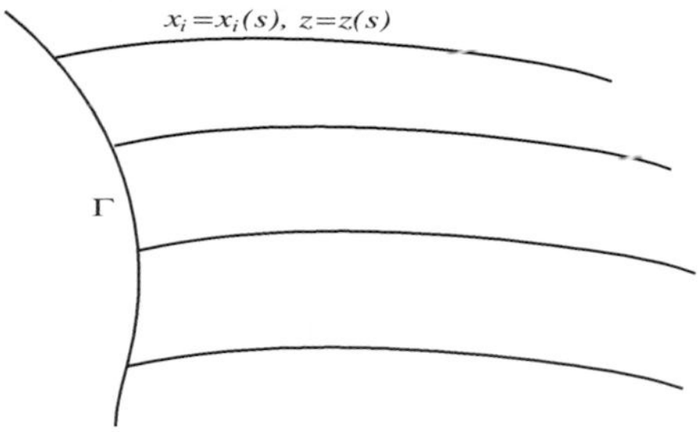

Fig. 5.3 The integral manifold is filled by the characteristic curves emanating from Γ

Finally, we will eliminate the last degrees of freedom by imposing the compatibility conditions

$$\sum_{i=1}^{n} p_i^0 \frac{\partial \varphi_i(u)}{\partial u_j}(\boldsymbol{u}) = \frac{\partial \varphi}{\partial u_j}(\boldsymbol{u}), \quad \forall \boldsymbol{u} \in U, \ j = 1, \dots, n-1. \tag{5.76}$$

The geometric significance of equations (5.76) should be clear: since the collection of vectors

$$\begin{bmatrix} \frac{\partial \varphi_1}{\partial u_j} \\ \vdots \\ \frac{\partial \varphi_n}{\partial u_j} \\ \frac{\partial \varphi}{\partial u_j} \end{bmatrix}, \quad j = 1, \dots, n-1,$$

forms a basis of the tangent space of the manifold Γ at $(\varphi_1(\boldsymbol{u}), \dots, \varphi_n(\boldsymbol{u}), z(\boldsymbol{u}))$, conditions (5.76) state that the vector

$$\begin{bmatrix} p_1^0 \\ \vdots \\ p_n^0 \\ -1 \end{bmatrix}$$

is normal to Γ at that point.

Taking into account the nondegeneracy assumption (5.71), we deduce from the implicit function theorem that the system (5.75), (5.76) determines uniquely the system of C^1-functions

$$p_i^0 = p_i^0(\boldsymbol{u}), \quad \boldsymbol{u} \in U_0 \subset U, \quad i = 1, \dots, n, \tag{5.77}$$

defined on some open subset $U_0 \subset U$. Substituting equations (5.74) and (5.77) into (5.73), we obtain for the functions x_i, z, p_i expressions of the type

$$\begin{aligned} x_i &= A_i(s, \boldsymbol{u}), \quad i = 1, \dots, n, \\ z &= B(s, \boldsymbol{u}), \quad s \in I_0, \ \boldsymbol{u} \in U_0, \\ p_i &= E_i(s, \boldsymbol{u}), \quad i = 1, \dots, n, \end{aligned} \tag{5.78}$$

where A_i, B and C_1 are C^1-functions on the domain $I_0 \times U_0 \subset I \times U$.

We will show that the first $(n+1)$ equations in (5.78) define parametrically a solution $z = z(\boldsymbol{x})$ of the Cauchy problem (5.69), (5.70) and the vector field $(E_1, \dots, E_n, -1)$ is normal to the graph of this function, more precisely,

$$E_i(s, \boldsymbol{u}) = z_{x_i}(s, \boldsymbol{u}), \ \ \forall (s, \boldsymbol{u}) \in I_0 \times U_0, \ \ i = 1, \ldots, n. \tag{5.79}$$

Indeed, from equations (5.70) and (5.71), we deduce

$$\frac{D(A_1, \ldots, A_n)}{D(u_1, \ldots, u_{n-1}, s)} = \frac{D(\varphi_1, \ldots, \varphi_n)}{D(u_1, \ldots, u_{n-1}, s)} \neq 0,$$

for $s = s_0$. We deduce that the system formed by the first n equations in (5.78) can be solved uniquely for $(s, \boldsymbol{u})$ in terms of $\boldsymbol{x}$. Substituting the resulting functions $s = s(\boldsymbol{x})$, $\boldsymbol{u} = \boldsymbol{u}(\boldsymbol{x})$ into the definition of z, we obtain a function $z = z(\boldsymbol{x})$.

To prove equalities (5.79), we start from the obvious equalities

$$B_s = \sum_{i=1}^{n} z_{x_i} \frac{\partial x_i}{\partial s} = \sum_{i=1}^{n} z_{x_i} \frac{\partial A_i}{\partial s}, \ \ \boldsymbol{u} \in U_0, \ \ s \in I_0,$$

$$B_{u_j} = \sum_{i=1}^{n} z_{x_i} \frac{\partial A_i}{\partial u_j}, \ \ j = 1, \ldots, n-1, \ \ \boldsymbol{u} \in U_0, \ \ s \in I_0.$$

Hence, equalities (5.1) are equivalent to the following system of equations

$$B_s = \sum_{i=1}^{n} E_i (A_i)_s, \ \ \boldsymbol{u} \in U_0, \ \ s \in I_0, \tag{5.80}$$

$$B_{u_j} = \sum_{i=1}^{n} E_i (A_i)_{u_j}, \ \ \boldsymbol{u} \in U_0, \ \ s \in I_0, \ \ j = 1, \ldots, n-1. \tag{5.81}$$

Equation (5.80) follows immediately from the characteristics equation (5.72).

To prove (5.81), we introduce the functions $L_j : I_0 \times U_0 \to \mathbb{R}$,

$$L_j(s, u) = \sum_{i=1}^{n} E_i (A_i)_{u_j} - B_{u_j}, \ \ j = 1, \ldots, n-1. \tag{5.82}$$

From equations (5.76), we deduce that

$$L_j(s_0, \boldsymbol{u}) = 0, \ \ \forall \boldsymbol{u} \in U_0, \ \ j = 1, \ldots, n-1. \tag{5.83}$$

On the other hand, from equalities (5.72), (5.78) and (5.82), we deduce that

$$\begin{aligned} \frac{\partial L_j}{\partial s} &= \sum_{i=1}^{n} (E_i)_s (A_i)_{u_j} + \sum_{i=1}^{n} E_i \frac{\partial^2 A_i}{\partial u_j \partial s} - \frac{\partial^2 B}{\partial u_j \partial s} \\ &= -\sum_{i=1}^{n} (E_i Z + X_i)(A_i)_{u_j} + \sum_{i=1}^{n} E_i (P_i)_{u_j} - \sum_{i=1}^{n} (E_i P_i)_{u_j}. \end{aligned} \tag{5.84}$$

Let us now observe that, for any $(s, \boldsymbol{u}) \in I_0 \times U_0$, we have the equality

$$F\big(A_1(s,\boldsymbol{u}), \dots, A_n(s,\boldsymbol{u}), B(s,\boldsymbol{u}), F_1(s,\boldsymbol{u}), \dots, F_n(s,\boldsymbol{u})\big) = 0. \tag{5.85}$$

Indeed, for $s = s_0$, equality (5.85) reduces to (5.75). On the other hand, using again system (5.72), we deduce that

$$\begin{aligned} &\frac{\partial}{\partial s} F\big(A_1(s,\boldsymbol{u}), \dots, A_n(s,\boldsymbol{u}), B(s,\boldsymbol{u}), F_1(s,\boldsymbol{u}), \dots, F_n(s,\boldsymbol{u})\big) \\ &= \sum_{i=1}^n X_i(A_i)_s - \sum_{i=1}^n P_i(E_i Z + X_i) + Z \sum_{i=1}^n E_i P_i = 0, \;\; \forall (s,\boldsymbol{u}) \in I_0 \times U_0, \end{aligned}$$

which immediately implies (5.85).

Differentiating (5.85) with respect to u_j, we deduce that

$$\sum_{i=1}^n X_i(A_i)_{u_j} + Z B_{u_j} + \sum_{i=1}^n P_i(E_i)_{u_j} = 0.$$

Using the above equality in (5.84), we find that

$$\frac{\partial L_j}{\partial s} = \left(B_{u_j} - \sum_{i=1}^n E_i(A_i)_{u_j} \right) Z = -L_s(s,\boldsymbol{u}) Z(s,\boldsymbol{u}),$$
$$\forall j = 1, \dots, n-1, \;\; (s,\boldsymbol{u}) \in I_0 \times U_0.$$

The above equality defines a first-order linear ODE (in the variable s) for each fixed $\boldsymbol{u} \in U_0$. For $s = s_0$, we have

$$L_j(s_0, \boldsymbol{u}) = 0,$$

and we obtain that

$$L_j(s, \boldsymbol{u}) = 0, \;\; \forall (s,\boldsymbol{u}) \in I_0 \times U_0, \;\; j = 1, \dots, n-1.$$

This proves equalities (5.81) and thus $z = z(\boldsymbol{x})$ is a solution of (5.69). Equality (5.85) shows that the graph of this function contains the manifold Γ.

The uniqueness of the solution of the Cauchy problem can be proved using the same method employed in the proof of Theorem 5.6. This completes the proof of Theorem 5.7. □

Remark 5.2 In the case $n = 2$, the Cauchy problem specializes to the following:

Find the surface $z = z(x, y)$ that satisfies the nonlinear first-order PDE

$$F(x, y, x, p, q) = 0, \tag{5.86}$$

and contains the curve

$$x = \varphi(t), \quad y = \psi(t), \quad z = \chi(t), t \in I, \tag{5.87}$$

where F is a C^2-function defined on a domain $\Omega \subset \mathbb{R}^5$, and φ, ψ, χ are C^1-functions on a real interval I.

If we denote by X, Y, Z, P, Q the functions

$$X = F_x, \quad Y = F_y, \quad Z = F_z, \quad P = F_p, \quad Q = F_q,$$

then the characteristic system (5.72) becomes

$$\begin{aligned} \frac{dx}{ds} = P, \quad \frac{dy}{ds} = Q, \quad \frac{dz}{ds} = pP + qQ, \\ \frac{dp}{ds} = -(pZ + X), \quad \frac{dq}{ds} = -(qZ + Y), \quad s \in I. \end{aligned} \tag{5.88}$$

The sought-after surface will be spanned by a one-parameter family of characteristic curves, that is, solutions of (5.88)

$$\begin{aligned} x &= x(s; s_0, y_0, z_0, p_0, q_0), \\ y &= y(s; s_0, y_0, z_0, p_0, q_0), \\ z &= z(s; s_0, y_0, z_0, p_0, q_0), \quad s \in I_0, \\ p &= p(s; s_0, y_0, z_0, p_0, q_0), \\ q &= q(s; s_0, y_0, z_0, p_0, q_0), \end{aligned} \tag{5.89}$$

where, according to the general procedure (see Eqs. (5.74), (5.75), (5.76)), the initial vector $(x_0, y_0, z_0, p_0, q_0)$ is determined by the conditions

$$\begin{aligned} x_0 = \varphi(t), \quad y_0 = \psi(t), \quad z_0 = \chi(t), \quad t \in I, \\ F(x_0, y_0, z_0, p_0, q_0) = 0, \\ p_0\varphi'(t) + q_0\psi'(t) = \chi'(t). \end{aligned}$$

In this fashion, system (5.89) defines the sought-after surface in parametric form

$$x = A(s, t), \quad y = B(s, t), \quad z = C(s, t).$$

Let us observe that, in this case, condition (5.71) becomes

$$\det \begin{bmatrix} P\big(\varphi(t), \psi(t), \chi(t)\big) & Q\big(\varphi(t), \psi(t), \chi(t)\big) \\ \varphi'(t) & \psi'(t) \end{bmatrix} \neq 0.$$

Example 5.7 Let us illustrate the above techniques on an optimal-time problem

$$T(\boldsymbol{x}_0) := \inf\{T; \;\; \boldsymbol{x}(T) = \boldsymbol{x}_0\}, \tag{5.90}$$

where $\boldsymbol{x}_0$ is an arbitrary point in $\mathbb{R}^n$ and the infimum is taken over all the solutions

$$\boldsymbol{x} \in C^1\big([0,T],\mathbb{R}^n\big)$$

of the differential inclusion

$$\begin{aligned} \boldsymbol{x}'(t) \in U\big(\boldsymbol{x}(t)\big),\ \ t \in [0,T], \\ \boldsymbol{x}(0) = 0. \end{aligned} \tag{5.91}$$

Above, for any $\boldsymbol{x}$, $U(\boldsymbol{x})$ is a compact subset of $\mathbb{R}^n$. In the sequel, we assume that the function $T : \mathbb{R}^n \to \mathbb{R}$ is C^1 on $\mathbb{R}^n \setminus \{0\}$.

If $\boldsymbol{x}^* \in C^1([0,T^*],\mathbb{R}^n)$ is an optimal arc in problem (5.90), that is, a solution of the multivalued system of ODEs (5.91) such that the infimum in (5.90) is attained for this solution, then we obviously have

$$T\big(\boldsymbol{x}^*(t)\big) = t,\ \ \forall t \in [0,T^*],$$

because T^* is optimal. Hence

$$\big(\nabla T^*(\boldsymbol{x}^*(t)\,), \boldsymbol{x}^{*\prime}(t)\,\big) = 1,\ \ \forall t \in [0,T^*].$$

In particular, for $t = T^*$, we have

$$\big(\nabla T(\boldsymbol{x}_0), \boldsymbol{u}_0\,\big) = 1. \tag{5.92}$$

On the other hand, for any solution of system (5.91) satisfying $x(T) = \boldsymbol{x}_0$, we have

$$T(\boldsymbol{x}(t)) \leq t - s + T(\boldsymbol{x}(s)),\ \ 0 \leq s \leq t \leq T.$$

Hence

$$(\nabla T(\boldsymbol{x}(t)), \boldsymbol{x}'(t)) \leq 1,\ \ \forall t \in [0,T].$$

It follows that, for $t = T$, we have

$$(\nabla T(\boldsymbol{x}_0), \boldsymbol{u}) \leq 1,\ \ \forall \boldsymbol{u} \in U(\boldsymbol{x}_0). \tag{5.93}$$

We set $z(\boldsymbol{x}_0) = T(\boldsymbol{x}_0)$ and we define the function $H_0 : \mathbb{R}^n \times \mathbb{R}^n \to \mathbb{R}$

$$H_0(\boldsymbol{x},\boldsymbol{w}) = \sup\big\{\,(\boldsymbol{w},\boldsymbol{u});\ \ \boldsymbol{u} \in U(\boldsymbol{x})\,\big\}.$$

From equalities (5.92) and (5.93), we deduce that z is a solution of the equation

$$H_0(\boldsymbol{x}, \nabla z(\boldsymbol{x})\,) = 1,\ \ \boldsymbol{x} \neq 0. \tag{5.94}$$

Consider the special case in which

$$U(\boldsymbol{x}) := \big\{ \boldsymbol{u} \in \mathbb{R}^n;\ \ \|\boldsymbol{u}\|_{\mathbf{e}} \le v(\boldsymbol{x}) \big\},\ \ \boldsymbol{x} \in \mathbb{R}^n,$$

where $v : \mathbb{R}^n \to (0, \infty)$ is a given C^1-function. In this case, we have

$$H_0(\boldsymbol{x}, \boldsymbol{w}) = \|\boldsymbol{w}\|_{\mathbf{e}} v(\boldsymbol{x}),$$

and so, Eq. (5.94) becomes

$$\|\nabla z(\boldsymbol{x})\|_{\mathbf{e}}^2 = \frac{1}{v(\boldsymbol{x})^2},\quad \forall x \neq 0. \tag{5.95}$$

In the special case where $n = 3$ and $v(\boldsymbol{x})$ is the speed of light in a nonhomogeneous medium, (5.95) is the fundamental equation of geometric optics and it is called the *eikonal equation*.

From Fermat's principle, it follows that the surface $z(x_1, x_2, x_3) = \lambda$ is the wave front of light propagation in a nonhomogeneous medium.

Equation (5.95), that is,

$$z_{x_1}^2 + z_{x_2}^2 + z_{x_3}^2 = n(\boldsymbol{x}) := v(\boldsymbol{x})^{-2},$$

has the form (5.69) and can be solved via the method described above. The solutions $x_i = x_i(s)$ to the characteristic system associated with (5.95),

$$\begin{aligned} \frac{dx_i}{ds} &= 2\pi,\ \ i = 1, 2, 3,\ \ \frac{dz}{ds} = 2(p_1^2 + p_2^2 + p_3^2) = 2n \\ \frac{dp_1}{ds} &= -n_{x_1},\ \ \frac{dp_3}{ds} = -n_{x_3},\ \ \frac{dp_3}{ds} = -n_{x_3}, \end{aligned}$$

represent, in the geometric optics context, light rays. Consider in the space $\mathbb{R}^4$ the surface Γ described by the parametrization

$$\begin{aligned} &x_1 = \varphi_1(u_1, u_2),\ \ x_2 = \varphi_2(u_1, u_2), \\ &x_3 = \varphi_3(u_1, u_2),\ z = \varphi(u_1, u_2). \end{aligned} \tag{5.96}$$

We want to solve the Cauchy problem associated with equation (5.94) and the surface Γ. We assume, for simplicity, that the medium is homogeneous, that is, n is a constant function. The characteristic curves of the equations (5.95) are given by

$$\begin{aligned} x_1 &= x_1^0 + 2p_1^0 s,\quad z = v^0 + 2ns, \\ x_2 &= x_2^0 + 2p_2^0 s,\quad p_1 = p_1^0,\ \ p_2 = p_2^0, \\ x_3 &= x_3^0 + p_3^0 s,\quad\ \ p_3 = p_3^0, \end{aligned} \tag{5.97}$$

and the initial conditions at $s = 0$ are determined by constraints (5.74), (5.75), (5.76), that is,

$$x_1^0 = \varphi_1(u_1, u_2), \quad x_2^0 = \varphi_2(u_1, u_2), \quad x_3^0 = \varphi_3(u_1, u_2),$$
$$z^0 = \varphi(u_1, u_2),$$
$$(p_1^0)^2 + (p_2^0)^2 + (p_3^0)^2 = n^2,$$
$$p_1^0 \frac{\partial \varphi_1}{\partial u_1} + p_2^0 \frac{\partial \varphi_2}{\partial u_1} + p_3^0 \frac{\partial \varphi_3}{\partial u_1} = \frac{\partial \varphi}{\partial u_1}, \quad p_1^0 \frac{\partial \varphi_1}{\partial u_2} + p_2^0 \frac{\partial \varphi_2}{\partial u_2} + p_3^0 \frac{\partial \varphi_3}{\partial u_2} = \frac{\partial \varphi}{\partial u_2},$$

from which we obtain a parametrization of the graph of the solution $z = z(x_1, x_2, x_3)$, which contains the manifold Γ.

5.6 Hamilton–Jacobi Equations

This section is devoted to the following problem

$$z_t + H_0(\boldsymbol{x}, z_{\boldsymbol{x}}, t) = 0, \quad t \in I \subset \mathbb{R}, \quad \boldsymbol{x} \in D \subset \mathbb{R}^{n-1}, \tag{5.98}$$
$$z(\boldsymbol{x}, 0) = \varphi(\boldsymbol{x}), \quad \forall \boldsymbol{x} \in D, \tag{5.99}$$

where $\boldsymbol{x} = (x_1, \ldots, x_{n-1})$, $z_{\boldsymbol{x}} = (z_{x_1}, \ldots, z_{x_{n-1}})$, H_0 is a C^2-function on a domain $\Omega \subset \mathbb{R}^{2n-1}$, and $\varphi \in C^1(D)$.

Problem (5.98), (5.99) is of type (5.69), (5.70), where the variable x_n was denoted by t and

$$F(x_1, ..., x_{n-1}, t, z, p_1, ..., p_{n-1}, p_n) = p_n + H_0(x_1, ..., x_{n-1}, p_1, ..., p_{n-1}, t),$$
$$\varphi_i(u_1, ..., u_{n-1}) = u_i, \quad \forall i = 1, ..., n-1.$$

We deduce by Theorem 5.7 that problem (5.98), (5.99) has a unique (locally defined) solution that can be determined using the strategy presented in the previous section.

Equation (5.98) is called the *Hamilton–Jacobi equation* and occupies a special place amongst the equations of mathematical physics being, among many other things, the fundamental equation of analytical mechanics. This equation appears in many other contexts, more often variational, which will be briefly described below.

Consider the C^1-function

$$L : \mathbb{R}^{n-1} \times \mathbb{R}^{n-1} \times \mathbb{R} \to \mathbb{R}$$

and define

$$H : \mathbb{R}^{n-1} \times \mathbb{R}^{n-1} \times \mathbb{R} \to \mathbb{R}$$

by setting

$$H(\boldsymbol{x}, \boldsymbol{p}, t) := \sup_{\boldsymbol{v} \in \mathbb{R}^{n-1}} \Big[(\boldsymbol{p}, \boldsymbol{v}) - L(\boldsymbol{x}, \boldsymbol{v}, t) \Big], \tag{5.100}$$

where $(-,-)$ denotes, as usual, the canonical scalar product on $\mathbb{R}^{n-1}$. In the sequel, we make the following assumption.

($\mathbf{A}_1$) *The function H is C^2 on a domain $\Omega \subset \mathbb{R}^{2n-1}$.*

If we interpret the function L at time t as a Lagrangian on the space $\mathbb{R}^{n-1} \times \mathbb{R}^{n-1}$, then the function H is, for any t, the corresponding Hamiltonian function (see (5.18)).

Consider the solution $w = w(\boldsymbol{x}, t)$ of the equation

$$w_t - H(\boldsymbol{x}, -w_{\boldsymbol{x}}, t) = 0, \quad \boldsymbol{x} \in \mathbb{R}^{n-1}, \quad t \in [0, T], \tag{5.101}$$

satisfying the Cauchy problem

$$w(\boldsymbol{x}, T) = \varphi(\boldsymbol{x}), \quad \boldsymbol{x} \in \mathbb{R}^{n-1}. \tag{5.102}$$

Using the change of variables

$$w(\boldsymbol{x}, t) = z(\boldsymbol{x}, T - t),$$

we see that equation (5.101) reduces to (5.98) with

$$H_0(\boldsymbol{x}, \boldsymbol{p}, t) = -H(\boldsymbol{x}, -\boldsymbol{p}, T - t).$$

Thus, we see that (5.101), (5.102) are equivalent to problem (5.98), (5.99).

Consider the following variational problem

$$\inf\left\{ \int_0^T L\big(\boldsymbol{x}(s), \boldsymbol{x}'(s), t\big) ds; \quad \boldsymbol{x} \in C^1([0, T], \mathbb{R}^{n-1}), \quad \boldsymbol{x}(0) = \boldsymbol{x}_0 \right\}, \tag{5.103}$$

and the function

$$S : \mathbb{R}^{n-1} \times [0, T] \to \mathbb{R},$$

given by

$$\begin{aligned} S(\boldsymbol{x}_0, t) = \inf\bigg\{ &\int_t^T L(\boldsymbol{x}(s), \boldsymbol{x}'(s), t) ds + \varphi(\boldsymbol{x}(T)); \\ &\boldsymbol{x} \in C^1([0, T], \mathbb{R}^{n-1}), \ \boldsymbol{x}(t) = \boldsymbol{x}_0 \bigg\}. \end{aligned} \tag{5.104}$$

In particular, $S(\boldsymbol{x}_0, 0)$ coincides with the infimum in (5.103). In analytical mechanics, the function S is called the *action functional* of the Lagrangian. A function $\boldsymbol{x}$ that realizes the infimum in (5.103) or (5.104) is called an *optimal arc*.

The connection between equation (5.101) and the infimum (5.104) will be described in Theorem 5.8 below. With this in mind, we consider the solution $w = w(\boldsymbol{x}, t)$ of problem (5.101), (5.102). We denote by $U(\boldsymbol{x}, t)$ the vector where the supremum in (5.100) is attained in the case $\boldsymbol{p} = -w_{\boldsymbol{x}}$, that is,

$$H\big(\boldsymbol{x}, -w_{\boldsymbol{x}}(\boldsymbol{x}, t)\big) = -\big(w_{\boldsymbol{x}}(\boldsymbol{x}, t), U(\boldsymbol{x}, t)\big) - L\big(\boldsymbol{x}, U(\boldsymbol{x}, t), t\big). \tag{5.105}$$

Concerning the function $U : \mathbb{R}^{n-1} \times [0, T] \to \mathbb{R}^{n-1}$, we will make the following assumption.
($\mathbf{A}_2$) *The function U is continuous and, for any $(\boldsymbol{x}_0, t) \in \mathbb{R}^{n-1} \times [0, T]$, the Cauchy problem*

$$\begin{aligned} \boldsymbol{x}'(s) &= U(\boldsymbol{x}(s), s), \quad t \leq s \leq T, \\ \boldsymbol{x}(t) &= \boldsymbol{x}_0, \end{aligned} \tag{5.106}$$

has a unique C^1-solution on the interval $[t, T]$.

Theorem 5.8 *Under the above assumptions, let $w : \mathbb{R}^{n-1} \times [0, T]$ be the solution of (5.101), (5.102). Then*

$$S(\boldsymbol{x}_0, t) = w(\boldsymbol{x}_0, t), \quad \forall \boldsymbol{x} \in \mathbb{R}^{n-1}, \quad t \in [0, T], \tag{5.107}$$

and, for any $(\boldsymbol{x}_0, t) \in \mathbb{R}^{n-1} \times [0, t]$, the solution $\boldsymbol{x} = \tilde{\boldsymbol{x}}(s)$ to problem (5.106) is an optimal arc for problem (5.104).

Proof Let $\boldsymbol{x} \in C^1\big([t, T], \mathbb{R}^{n-1}\big)$ be an arbitrary function such that $\boldsymbol{x}(t) = \boldsymbol{x}_0$. From the obvious equality

$$\frac{d}{ds} w\big(\boldsymbol{x}(s), s\big) = w_s\big(\boldsymbol{x}(s), s\big) + \big(w_{\boldsymbol{x}}\big(\boldsymbol{x}(s), s\big), \boldsymbol{x}'(s)\big) \;\; \forall s \in [t, T], \tag{5.108}$$

and from (5.100), (5.101), it follows that

$$\begin{aligned} \frac{d}{ds} w\big(\boldsymbol{x}(s), s\big) &= H\big(\boldsymbol{x}(s), -w_{\boldsymbol{x}}\big(\boldsymbol{x}(s), s\big), s\big) + \big(w_{\boldsymbol{x}}\big(\boldsymbol{x}(s), s\big), \boldsymbol{x}'(s)\big) \\ &\geq L\big(\boldsymbol{x}(s), \boldsymbol{x}'(s), s\big), \quad \forall s \in [t, T]. \end{aligned}$$

Hence

$$w\big(\boldsymbol{x}_0, t\big) \leq \int_t^T L\big(\boldsymbol{x}(s), \boldsymbol{x}'(s), s\big) ds + \varphi\big(\boldsymbol{x}(T)\big).$$

Since the function $\boldsymbol{x}$ is arbitrary, we deduce that

$$w(\boldsymbol{x}_0, t) \leq S(\boldsymbol{x}_0, t). \tag{5.109}$$

Consider now the solution $\boldsymbol{x} = \tilde{\boldsymbol{x}}(t)$ of the Cauchy problem (5.106). From equality (5.105), it follows that

$$\begin{aligned} \frac{d}{ds} w\big(\tilde{\boldsymbol{x}}(s), s\big) &= H\big(\tilde{\boldsymbol{x}}(s), -w_{\boldsymbol{x}}\big(\tilde{\boldsymbol{x}}(s), s\big), s\big) + \big(w_{\boldsymbol{x}}\big(\tilde{\boldsymbol{x}}(s), s\big), \tilde{\boldsymbol{x}}'(s)\big) \\ &= L\big(\tilde{\boldsymbol{x}}(s), U(\tilde{\boldsymbol{x}}(s), s), s\big), \quad \forall s \in [t, T]. \end{aligned}$$

Integrating over the interval $[t, T]$, we deduce that

$$w(\boldsymbol{x}_0, t) = \int_t^T L\big(\tilde{\boldsymbol{x}}(s), U(\tilde{\boldsymbol{x}}(s), s), s\big)ds + \varphi\big(\tilde{\boldsymbol{x}}(T)\big) \geq S(\boldsymbol{x}_0, T).$$

The last inequality coupled with (5.109) shows that $w \equiv S$. On the other hand, from the last inequality we also deduce that the solution $\boldsymbol{x}$ of problem (5.106) is also an optimal arc for problem (5.104). □

Let us now assume that, besides $(\mathbf{A}_1)$ and $(\mathbf{A}_2)$, we know that, for any $(\boldsymbol{x}, t) \in \mathbb{R}^{n-1} \times [0, T]$, the function $\boldsymbol{v} \mapsto L(\boldsymbol{x}, \boldsymbol{v}, t)$ is convex. According to Proposition A.2, we have

$$L(\boldsymbol{x}, \boldsymbol{v}, t) + H(\boldsymbol{x}, \boldsymbol{p}, t) = (\boldsymbol{p}, \boldsymbol{v}) \ \text{ for } \boldsymbol{v} = H_{\boldsymbol{p}}(\boldsymbol{x}, \boldsymbol{p}). \tag{5.110}$$

The function U from (5.105) is thus given by

$$U(\boldsymbol{x}, t) = H_{\boldsymbol{p}}\big(\boldsymbol{x}, -w_{\boldsymbol{x}}(\boldsymbol{x}, t)\big), \ \ \forall \boldsymbol{x} \in \mathbb{R}^{n-1}, \ \ t \in [0, T]. \tag{5.111}$$

In the particular case

$$L(\boldsymbol{x}, \boldsymbol{v}, t) = \frac{1}{2}\left(\big(Q(t)\boldsymbol{x}, \boldsymbol{x}\big) + \frac{1}{2}\|\boldsymbol{v}\|_{\mathbf{e}}^2\right), \ \ \forall (\boldsymbol{x}, \boldsymbol{v}, t) \in \mathbb{R}^{n-1} \times \mathbb{R}^{n-1} \times [0, T],$$

where $Q(t)$ is an $(n-1) \times (n-1)$ matrix, we have

$$H(\boldsymbol{x}, \boldsymbol{p}, t) = \frac{1}{2}\Big(\|\boldsymbol{p}\|_{\mathbf{e}}^2 - \big(Q(t)\boldsymbol{x}, \boldsymbol{x}\big)\Big), \ \ \forall (\boldsymbol{x}, \boldsymbol{p}, t) \in \mathbb{R}^{n-1} \times \mathbb{R}^{n-1} \times [0, T],$$

and we deduce that

$$U(\boldsymbol{x}, t) = -w_{\boldsymbol{x}}(\boldsymbol{x}, t), \ \ \forall (\boldsymbol{x}, t) \in \mathbb{R}^{n-1} \times \mathbb{R}.$$

Equation (5.101) becomes

$$w_t - \frac{1}{2}\|w_{\boldsymbol{x}}\|_{\mathbf{e}}^2 = -\frac{1}{2}\big(Q(t)\boldsymbol{x}, \boldsymbol{x}\big). \tag{5.112}$$

If we seek for (5.112) a solution of the form

$$w(\boldsymbol{x}, t) = \big(P(t)\boldsymbol{x}, \boldsymbol{x}\big), \ \ \boldsymbol{x} \in \mathbb{R}^{n-1}, \ \ t \in [0, T], \tag{5.113}$$

where $P(t)$ is an $(n-1) \times (n-1)$ real matrix, then (5.112) becomes

$$P'(t) - P(t)^2 = -Q(t), \ \ t \in [0, T]. \tag{5.114}$$

This is a Riccati equation of type (2.50) that we have investigated earlier. From the uniqueness of the Cauchy problem for (5.112), we deduce that any solution of (5.112) has the form (5.113).

Returning to the variational problem (5.103), and recalling that any trajectory of a mechanical system with $(n-1)$-degrees of freedom is an optimal arc for a system of type (5.103), we deduce from Theorem 5.8 that the Hamilton–Jacobi equation (5.101) represents, together with the Hamiltonian systems (5.21), another way of describing the motions in classical mechanics. There exists therefore an intimate relationship between the Hamiltonian systems

$$\begin{aligned} \boldsymbol{p}'(t) &= -\frac{\partial H}{\partial \boldsymbol{x}}\big(\boldsymbol{x}(t), \boldsymbol{p}(t)\big) \\ \boldsymbol{x}'(t) &= \frac{\partial H}{\partial \boldsymbol{p}}\big(\boldsymbol{x}(t),\ \boldsymbol{p}(t)\big), \quad t \in [0, T], \end{aligned} \tag{5.115}$$

and the Hamilton–Jacobi equation (5.101).

In the case of the one-dimensional motion of (n)-particles of masses $m_1, \dots, m_{n-1}$, the Hamiltonian is given by (5.22), that is,

$$H(x_1, \dots, x_n; p_1, \dots, p_n) = \sum_{k=1}^{n} \frac{1}{2m_k} p_k^2 + V(x_1, \dots, x_n), \tag{5.116}$$

and the corresponding Hamilton–Jacobi equation becomes

$$W_t(\boldsymbol{x}, t) - \sum_{i=1}^{n} \frac{1}{2m_i} W_{x_i}^2(\boldsymbol{x}, t) = V(\boldsymbol{x}),\ \boldsymbol{x} \in \mathbb{R}^n,\quad t \in [0, T]. \tag{5.117}$$

In quantum mechanics, the motion of a particle is defined by its wave function $\psi(t, \boldsymbol{x})$, which has the following significance: the integral

$$\int_E |\psi(\boldsymbol{x}, t)|^2 d\boldsymbol{x}$$

represents the probability that the particle is located at time t in the region $E \subset \mathbb{R}^3$.

In classical mechanics, the Hamiltonian of a particle of mass m is given by

$$H(x, p) = \frac{1}{2m} p^2 + V(x), \quad (x, p) \in \mathbb{R}^2, \tag{5.118}$$

where p is the momentum of the particle during the motion. In quantum mechanics, the momentum of the particle is represented by the differential operator

$$\mathbf{p} := -i\hbar \frac{\partial}{\partial x},$$

where $\hbar$ is Planck's constant.

Using the analogy with classical mechanics, the quantum Hamiltonian is defined to be the differential operator

$$\mathbf{H} = \frac{1}{2m}\mathbf{p}\cdot\mathbf{p} + V(x) = -\frac{\hbar^2}{2m}\frac{\partial^2}{\partial x^2} + V(x),$$

and, formally, the Hamilton–Jacobi equation becomes

$$i\hbar\psi_t(x,t) + \frac{\hbar^2}{2m}\frac{\partial^2\psi}{\partial x^2} - V(x)\psi = 0,$$

or, equivalently,

$$\psi_t(x,t) - \frac{i\hbar}{2m}\psi_{xx}(x,t) + \frac{i}{\hbar}V(x)\psi(x,t) = 0. \tag{5.119}$$

Equation (5.119) satisfied by the wave function Ψ is called *Schrödinger's equation*. It is the fundamental equation of quantum mechanics and we have included it here to highlight its similarities with the Hamilton–Jacobi equation in classical mechanics.

Remark 5.3 In general, the Hamilton–Jacobi equation (5.98)–(5.99) does not have a global C^1-solution, and the best one can obtain for this problem is a weak or generalized solution. One such concept of solution for which one has existence and uniqueness under certain continuity and growth conditions on the Hamiltonian function H is that of the *viscosity solution* introduced by M.G. Crandall and P.L. Lions [8].

Problems

5.1 Determine two independent, prime integrals of the differential system

$$x_1' = x_2^2, \;\; x_2' = x_2x_3, \;\; x_3' = -x_2^2. \tag{5.120}$$

Hint. Use Theorem 5.2 and the construction given in its proof.

5.2 The differential system

$$\begin{aligned} I_1x_1' &= (I_2 - I_1)x_2x_3,\\ I_2x_2' &= (I_3 - I_1)x_3x_1,\\ I_3x_3' &= (I_1 - I_2)x_1x_2, \end{aligned} \tag{5.121}$$

describes the motion of a rigid body with a fixed point. Find a prime integral of this system.
Hint. Check equation (5.2) for $U(x_1, x_2, x_3) = I_1x_1^2 + I_2x_2^2 + I_3x_3^2$.

5.3 Find the general solution of the linear, first-order PDE

$$z_x + \alpha z_y = 0. \tag{5.122}$$

5.4 Find the integral surface of the equation

$$(x - z)z_x + (y - z)z_y = 2z, \tag{5.123}$$

with the property that it contains the curve

$$\Gamma := \{x - y = 2, \quad x + z = 1\} \subset \mathbb{R}^3.$$

Hint. Use the method described in Sect. 5.3.

5.5 Let A be an $n \times n$ real matrix and

$$f : \mathbb{R}^n \times \mathbb{R} \to \mathbb{R}, \quad \varphi : \mathbb{R}^n \to \mathbb{R},$$

be C^1-functions. Prove that the solution of the Cauchy problem

$$\begin{aligned} z_t(\boldsymbol{x}, t) - \big(A\boldsymbol{x}, z_x(\boldsymbol{x}, t)\big) &= f(\boldsymbol{x}, t), \quad \boldsymbol{x} \in \mathbb{R}^n, \quad t \in \mathbb{R}, \\ z(\boldsymbol{x}, 0) &= \varphi(\boldsymbol{x}), \qquad \boldsymbol{x} \in \mathbb{R}^n, \end{aligned} \tag{5.124}$$

is given by the formula

$$z(\boldsymbol{x}, t) = \varphi\big(e^{tA}\boldsymbol{x}\big) + \int_0^t f\big(e^{(t-s)A}\boldsymbol{x}, s\big)ds, \quad \forall (\boldsymbol{x}, t) \in \mathbb{R}^n \times \mathbb{R}. \tag{5.125}$$

5.6 Let A be an $n \times n$ real matrix. Using the successive approximations method, prove the existence and uniqueness of the solution of the Cauchy problem

$$\begin{aligned} z_t(\boldsymbol{x}, t) - \big(A\boldsymbol{x}, z_z(\boldsymbol{x}, t)\big) &= F\big(\boldsymbol{x}, t, z(\boldsymbol{x}, t)\big), \\ z(\boldsymbol{x}, 0) &= \varphi(\boldsymbol{x}), \qquad (\boldsymbol{x}, t) \in \mathbb{R}^n \times \mathbb{R}, \end{aligned} \tag{5.126}$$

where $\varphi : \mathbb{R}^n \to \mathbb{R}$ and $F : \mathbb{R}^n \times \mathbb{R} \times \mathbb{R} \to \mathbb{R}$ are C^1-functions.
Hint. Using (5.125), we can transform (5.126) into an integral equation

$$z(\boldsymbol{x}, t) = \varphi\big(e^{tA}\boldsymbol{x}\big) + \int_0^t F\big(e^{(t-s)A}\boldsymbol{x}, s, z\big(e^{(t-s)A}\boldsymbol{x}, s\big)\big)ds,$$

which can be solved using the successive approximations method.

5.7 Prove that the solution $z = z(x_1, \dots, x_n, t)$ of the Cauchy problem

$$\begin{aligned} z_t + \sum_{i=1}^{n} a_i z_{x_i} &= f(\boldsymbol{x}, t), \quad \boldsymbol{x} = (x_1, \dots, x_n) \in \mathbb{R}^n, \quad t \in \mathbb{R}, \\ z(\boldsymbol{x}, 0) &= \varphi(\boldsymbol{x}), \quad \boldsymbol{x} \in \mathbb{R}^n, \end{aligned} \tag{5.127}$$

where a_i are real constants and

$$f : \mathbb{R}^n \times \mathbb{R} \to \mathbb{R}, \quad \varphi : \mathbb{R}^n \to \mathbb{R},$$

are C^1-functions, is given by the formula

$$z(\boldsymbol{x}, t) = \varphi(\boldsymbol{x} - t\boldsymbol{a}) + \int_0^t f\big(\boldsymbol{x} - (t-s)\boldsymbol{a}, s\big) ds, \quad \boldsymbol{a} := (a_1, \dots, a_n).$$

Hint. Use the method of characteristics in Sect. 5.3.

5.8 The equation

$$\frac{\partial n}{\partial t} + \sum_{i=1}^{3} v_i \frac{\partial n}{\partial x_i} = \int_{\mathbb{R}^3} K(\boldsymbol{x}, \boldsymbol{\theta}, \boldsymbol{v}) n(\boldsymbol{x}, \boldsymbol{\theta}, t) d\boldsymbol{\theta},$$

$$n(\boldsymbol{x}, \boldsymbol{v}, 0) = n_0(\boldsymbol{x}, \boldsymbol{v}), \quad \boldsymbol{x} = (x_1, x_2, x_3), \quad \boldsymbol{v} = (v_1, v_2, v_3),$$

is called the *Boltzman transport equation* and describes the motion of neutrons where $n = n(\boldsymbol{x}, \boldsymbol{v}, t)$ is the density of neutrons having velocity $v = v(\boldsymbol{x})$ at time t and at the point $\boldsymbol{x} \in \mathbb{R}^3$. The function

$$K : \mathbb{R}^3 \times \mathbb{R}^3 \times \mathbb{R}^3 \to \mathbb{R}$$

is continuous and absolutely integrable.

Using the result in the previous exercise, prove an existence and uniqueness result for the transport equation by relying on the successive approximations method or Banach's contraction principle (Theorem A.2).

5.9 Prove that, if the function φ is C^1 and increasing on $\mathbb{R}$, then the solution of the Cauchy problem

$$\begin{aligned} z_x + z z_y &= 0, \quad x \geq 0, \quad y \in \mathbb{R} \\ z(0, y) &= \varphi(y), \quad y \in \mathbb{R}, \end{aligned} \tag{5.128}$$

exists on the entire half-plane $\{(x, y) \in \mathbb{R}^2; \ x \geq 0\}$, and it is given by the formula

$$z(x, y) = \varphi\big((1 + x\varphi)^{-1}(y) \big),$$

where $(1 + x\varphi)^{-1}$ denotes the inverse of the function

$$t \in \mathbb{R} \mapsto t + x\varphi(t) \in \mathbb{R}.$$

Generalize this result to equations of type (5.50) and then use it to solve the Cauchy problem

$$\begin{aligned} w_x + \frac{1}{2}|w_y|^2 &= 0, \qquad x \geq 0, \;\; y \in \mathbb{R}, \\ w(0, y) &= \psi(y), \quad y \in \mathbb{R}, \end{aligned} \tag{5.129}$$

where $\psi : \mathbb{R} \to \mathbb{R}$ is a convex C^2-function.
Hint. Differentiating (5.129) with respect to y, we obtain an equation of type (5.128).

5.10 The Hamilton–Jacobi equation

$$h_t(x, t) + Q(h_x(x, t)) = 0, \;\; x \in \mathbb{R}, \;\; t \geq 0, \tag{5.130}$$

was used to model the process of mountain erosion, where $h(x, t)$ is the altitude of (a section of) the mountain at time t and at location x. Solve equation (5.130) with the Cauchy condition $h(x, 0) = 1 - x^2$ in the case $Q(u) = u^4$.

5.11 Extend Theorem 5.8 to the case when problem (5.103) is replaced by an optimal control problem of the form

$$\inf_{\boldsymbol{u}} \left\{ \int_0^T L(\boldsymbol{x}(s), \boldsymbol{u}(s), s)ds + \varphi(\boldsymbol{x}(T)) \right\}, \tag{5.131}$$

where

$$\begin{aligned} \boldsymbol{u} \in C\big([0, T]; \mathbb{R}^m\big), \;\; \boldsymbol{x} \in C^1\big([0, T]; \mathbb{R}^{n-1}\big), \\ \boldsymbol{x}'(s) = \boldsymbol{f}\big(s, \boldsymbol{x}(s), \boldsymbol{u}(s)\big), \;\; \boldsymbol{x}(0) = \boldsymbol{x}_0, \end{aligned}$$

and

$$\boldsymbol{f} : [0, T] \times \mathbb{R}^{n-1} \times \mathbb{R}^m \to \mathbb{R}^{n-1}$$

is a C^1-function. Determine the Hamilton–Jacobi equation associated with (5.131) in the special case when $m = n - 1$ and

$$L(\boldsymbol{x}, \boldsymbol{u}) = \frac{1}{2}\big((Q\boldsymbol{u}, \boldsymbol{u}) + \|\boldsymbol{u}\|_{\mathbf{e}}^2\big), \;\; \boldsymbol{f}(s, \boldsymbol{x}, \boldsymbol{u}) = A\boldsymbol{y} + \boldsymbol{u}, \tag{5.132}$$

where A, Q are $(n-1) \times (n-1)$ real matrices.
Hint. Theorem 5.8 continues to hold with the Hamiltonian function given by the formula

$$H(\boldsymbol{x}, \boldsymbol{p}, t) := \sup_{\boldsymbol{u} \in \mathbb{R}^m} \big(\boldsymbol{p}, \boldsymbol{f}(t, \boldsymbol{x}, \boldsymbol{u})\big) - L\big(\boldsymbol{x}, \boldsymbol{f}(t, \boldsymbol{x}, \boldsymbol{u}), t\big).$$

In the special case when L and f are defined by (5.132), the Hamiltonian–Jacobi equations have the form

$$w_t(\boldsymbol{x}, t) + \big(A\boldsymbol{x}, w_{\boldsymbol{x}}(\boldsymbol{x}, t)\big) - \frac{1}{2}\|w_{\boldsymbol{x}}(\boldsymbol{x}, t)\|_{\mathbf{e}}^2 = -\frac{1}{2}\big(Q\boldsymbol{x}, \boldsymbol{x}\big). \qquad (5.133)$$

Differentiating (5.133) we obtain the Riccati equation (see (5.114))

$$P'(t) + A^*P(t) + P(t)A - P^2(t) = -Q(t), \quad t \in [0, T]. \qquad (5.134)$$

Appendix

A.1 Finite-dimensional Normed Spaces

Let $\mathbb{R}^n$ be the standard real vector space of dimension n. Its elements are n-dimensional vectors of the form $\boldsymbol{x} = (x_1, \dots, x_n)$. Often we will represent $\boldsymbol{x}$ as a column vector. The real numbers $x_1, \dots, x_n$ are called the *coordinates* of the vector $\boldsymbol{x}$. The addition of vectors is given by coordinatewise addition, and the multiplication of a vector by a real scalar is defined analogously. The space $\mathbb{R}^1$ is identified with the real line $\mathbb{R}$.

To a real matrix of size $n \times m$ (n-rows, m-columns), we associate a linear map

$$\widetilde{B} : \mathbb{R}^m \to \mathbb{R}^n,$$

given by the formula $\widetilde{B}\boldsymbol{x} = B\boldsymbol{x}$, where $B\boldsymbol{x}$ denotes the column vector

$$B\boldsymbol{x} := \begin{bmatrix} \sum\limits_{j=1}^{m} b_{1j}x_j \\ \sum\limits_{j=1}^{m} b_{2j}x_j \\ \vdots \\ \sum\limits_{j=1}^{n} b_{mj}x_j \end{bmatrix},$$

where b_{ij} denotes the entry of B situated at the intersection of the i-th row with the j-th column, $1 \leq i \leq m$, $1 \leq j \leq n$. Conversely, any linear map $\mathbb{R}^m \to \mathbb{R}^n$ has this form.

The *adjoint* of B is the $m \times n$ matrix B^* with entries

$$b^*_{ji} := b_{ij}, \quad \forall 1 \leq i \leq n, \quad 1 \leq j \leq m.$$

V. Barbu, *Differential Equations*, Springer Undergraduate Mathematics Series,
DOI 10.1007/978-3-319-45261-6

In particular, any $n \times n$ real matrix A induces a linear self-map of $\mathbb{R}^m$. The matrix A is called *nonsingular* if its determinant is non-zero. In this case, there exists an inverse matrix, denoted by A^{-1} and uniquely determined by the requirements

$$A \cdot A^{-1} = A^{-1} \cdot A = \mathbb{1}_n,$$

where we denoted by $\mathbb{1}_n$ the identity $n \times n$ matrix.

Definition A.1 A *norm* on $\mathbb{R}^n$ is a real function on $\mathbb{R}^n$, usually denoted by $\| - \|$, and satisfying the following requirements.

(i) $\|\boldsymbol{x}\| \geq 0, \forall \boldsymbol{x} \in \mathbb{R}^n$.
(ii) $\|\boldsymbol{x}\| = 0$ if and only if $\boldsymbol{x} = 0$.
(iii) $\|\lambda \boldsymbol{x}\| = |\lambda| \, \|\boldsymbol{x}\|, \forall \lambda \in \mathbb{R}, \boldsymbol{x} \in \mathbb{R}^n$.
(iv) $\|\boldsymbol{x} + \boldsymbol{y}\| \leq \|\boldsymbol{x}\| + \|\boldsymbol{y}\|, \forall \boldsymbol{x}, \boldsymbol{y} \in \mathbb{R}^n$.

There exist infinitely many norms on the space $\mathbb{R}^n$. We leave the reader to verify that the following functions on $\mathbb{R}^n$ are norms.

$$\|\boldsymbol{x}\| := \max_{1 \leq i \leq n} |x_i|, \tag{A.1}$$

$$\|\boldsymbol{x}\|_1 = \sum_{i=1}^{n} |x_i|, \tag{A.2}$$

$$\|\boldsymbol{x}\|_{\mathbf{e}} := \left(\sum_{i=1}^{n} x_i^2 \right)^{\frac{1}{2}}. \tag{A.3}$$

Any norm on $\mathbb{R}^n$ induces a topology on $\mathbb{R}^n$ that allows us to define the concept of *convergence* and the notion of *open ball*. Thus the sequence $(\boldsymbol{x}^j)_{j \geq 1}$ in $\mathbb{R}^n$ converges to $\boldsymbol{x}$ in $\mathbb{R}^n$ in the norm $\| - \|$ if

$$\lim_{j \to \infty} \|\boldsymbol{x}^j - \boldsymbol{x}\| = 0.$$

The *open ball* of center $\boldsymbol{a} \in \mathbb{R}^n$ and radius r is the set

$$\left\{ \boldsymbol{x} \in \mathbb{R}^n; \quad \|\boldsymbol{x} - \boldsymbol{a}\| < a \right\}$$

and the *closed ball* of center $\boldsymbol{a}$ and radius r is the set

$$\left\{ \boldsymbol{x} \in \mathbb{R}^n; \quad \|\boldsymbol{x} - \boldsymbol{a}\| \leq a \right\}.$$

We can define the topological notions of closure, open and closed sets and continuity. Thus, a set $\mathbb{R}^n$ is called *open* if, for any point $\boldsymbol{x}_0 \in D$, there exists an open ball centered at $\boldsymbol{x}_0$ and contained in D. The set $C \subset \mathbb{R}^n$ is called *closed* if, for any convergent sequence of points in C, the limit is also a point in C. The set $K \subset \mathbb{R}^n$ is called *compact* if any sequence of points in K contains a subsequence that converges to a point in K. Finally, a set D is called *bounded* if it is contained in some ball. For a more detailed investigation of the space $\mathbb{R}^n$, we refer the reader to a textbook on real analysis in several variables. e.g., Lang 2005.

It is not to hard to see from the above that the convergence in any norm is equivalent to coordinatewise convergence. More precisely, this means that given a sequence $(\boldsymbol{x}^j)$ in $\mathbb{R}^n$, $\boldsymbol{x}^j = (x_1^j, \ldots, x_n^j)$, then

$$\lim_{j\to\infty} \boldsymbol{x}^j = \boldsymbol{x} = (x_1, \ldots, x_n) \Longleftrightarrow \lim_{j\to\infty} x_k^j = x_k, \quad \forall k = 1, \ldots, n.$$

This fact can be expressed briefly by saying that *any two norms on* $\mathbb{R}^n$ *are equivalent*. The next result phrases this fact in an equivalent form.

Lemma A.1 *Let* $\| - \|_1$ *and* $\| - \|_2$ *be two arbitrary norms on* $\mathbb{R}^n$. *Then there exists a constant* $C > 1$ *such that*

$$\frac{1}{C}\|\boldsymbol{x}\|_2 \leq \|\boldsymbol{x}\|_1 \leq C\|\boldsymbol{x}\|_2. \tag{A.4}$$

Thus, the notions of open, closed and compact subsets of $\mathbb{R}^n$ are the same for all the norms on $\mathbb{R}^n$.

Given a real $n \times n$ matrix with entries $\{a_{ij};\ 1 \leq i, j \leq n\}$, its norm is the real number

$$\|A\| = \max_i \sum_{j=1}^{n} |a_{ij}|. \tag{A.5}$$

In the vector space of real $n \times n$ matrices (which is a linear space of dimension n^2 and thus isomorphic to $\mathbb{R}^{n^2}$), the map $A \to \|A\|$ satisfies all the conditions (i)–(iv) in Definition A.1.

(i) $\|A\| \geq 0$.
(ii) $\|A\| = 0$ if and only if $A = 0$.
(iii) $\|\lambda A\| = |\lambda|\ \|A\|$, $\forall \lambda \in \mathbb{R}$.
(iv) $\|A + B\| \leq \|A\| + \|B\|$.

We will say that the sequence of matrices $(A_j)_{j\geq 1}$ converges to the matrix A as $j \to \infty$, and we will denote this by

$$A = \lim_{j\to\infty} A_j,$$

if

$$\lim_{j\to\infty} \|A_j - A\| = 0.$$

If we denote by $a_{k\ell}^j$, $1 \leq k, \ell \leq n$, the entries of the matrix A_j, and by $a_{k\ell}$, $1 \leq k, \ell \leq n$ the entries of the matrix A, then

$$\lim_{j\to\infty} A_j = A \Longleftrightarrow \lim_{j\to\infty} a^j_{k\ell} = a_{k\ell}, \quad \forall k, \ell.$$

Let us point out that if $\| - \|$ is the norm on $\mathbb{R}^n$ defined by (A.1), then we have the inequality

$$\|A\boldsymbol{x}\| \leq \|A\| \, \|\boldsymbol{x}\|, \quad \forall \boldsymbol{x} \in \mathbb{R}^n. \tag{A.6}$$

A.2 Euclidean Spaces and Symmetric Operators

Given two vectors

$$\boldsymbol{x} = (x_1, \dots, x_n), \quad \boldsymbol{y} = (y_1, \dots, y_n) \in \mathbb{R}^n,$$

we define their *scalar* or *inner product* to be the real number

$$(\boldsymbol{x}, \boldsymbol{y}) := \sum_{k=1}^{n} x_k y_k. \tag{A.7}$$

We view the scalar product as a function $(-, -) : \mathbb{R}^n \times \mathbb{R}^n \to \mathbb{R}$. It is not hard to see that it satisfies the following properties.

$$(\boldsymbol{x}, \boldsymbol{y}) = (\boldsymbol{y}, \boldsymbol{x}), \quad \forall \boldsymbol{x}, \boldsymbol{y} \in \mathbb{R}^n. \tag{A.8}$$

$$(\boldsymbol{x}, \boldsymbol{y} + \boldsymbol{z}) = (\boldsymbol{x}, \boldsymbol{y}) + (\boldsymbol{x}, \boldsymbol{z}), \; (\lambda \boldsymbol{x}, \boldsymbol{y}) = \lambda(\boldsymbol{x}, \boldsymbol{y}), \; \forall \boldsymbol{x}, \boldsymbol{y}, \boldsymbol{z} \in \mathbb{R}^n, \; \lambda \in \mathbb{R}, \tag{A.9}$$

$$(\boldsymbol{x}, \boldsymbol{x}) \geq 0, \quad \forall \boldsymbol{x} \in \mathbb{R}, \quad (\boldsymbol{x}, \boldsymbol{x}) = 0 \Longleftrightarrow \boldsymbol{x} = 0. \tag{A.10}$$

We observe that the function $\| - \|_{\mathbf{e}}$ defined by

$$\|\boldsymbol{x}\|_{\mathbf{e}} := (\boldsymbol{x}, \boldsymbol{x})^{\frac{1}{2}}, \quad \forall \boldsymbol{x} \in \mathbb{R}^n, \tag{A.11}$$

is precisely the norm in (A.3). The vector space $\mathbb{R}^n$ equipped with the scalar product (A.7) and the norm (A.11) is called the (real) n-dimensional Euclidean space.

In Euclidean spaces, one can define the concepts of orthogonality, symmetric operators and, in general, one can successfully extend a large part of classical Euclidean geometry. Here we will limit ourselves to presenting only a few elementary results.

Lemma A.2 *Let A be an $n \times n$ matrix and A^* its adjoint. Then*

$$(A\boldsymbol{x}, \boldsymbol{y}) = (\boldsymbol{x}, A^*\boldsymbol{y}), \quad \forall \boldsymbol{x}, \boldsymbol{y} \in \mathbb{R}^n. \tag{A.12}$$

The proof of the above result is by direct computation. In particular, we deduce that, if A is a symmetric matrix, $A = A^*$, then we have the equality

$$(A\boldsymbol{x}, \boldsymbol{y}) = (\boldsymbol{x}, A\boldsymbol{y}), \quad \forall \boldsymbol{x}, \boldsymbol{y} \in \mathbb{R}^n. \tag{A.13}$$

A real, symmetric $n \times n$ matrix P is called *positive definite* if

$$(P\boldsymbol{x}, \boldsymbol{x}) > 0, \quad \forall \boldsymbol{x} \in \mathbb{R}^n \setminus \{0\}. \tag{A.14}$$

The matrix P is called *positive* if

$$(P\boldsymbol{x}, \boldsymbol{x}) \geq 0, \quad \forall \boldsymbol{x} \in \mathbb{R}^n.$$

Lemma A.3 *A real $n \times n$ matrix is positive definite if and only if there exists a positive constant ω such that*

$$(P\boldsymbol{x}, \boldsymbol{x}) \geq \omega \|\boldsymbol{x}\|^2, \quad \forall \boldsymbol{x} \in \mathbb{R}^n. \tag{A.15}$$

Proof Obviously, condition (A.15) implies (A.14). To prove the converse, consider the set

$$M := \{\boldsymbol{x} \in \mathbb{R}^n; \quad \|\boldsymbol{x}\| = 1\},$$

and the function

$$q : M \to \mathbb{R}, \quad q(\boldsymbol{x}) = (P\boldsymbol{x}, \boldsymbol{x}).$$

The set M is compact and the function q is continuous and thus there exists an $\boldsymbol{x}_0 \in M$ such that

$$q(\boldsymbol{x}_0) \leq q(\boldsymbol{x}), \quad \forall \boldsymbol{x} \in M.$$

Condition (A.14) implies $q(\boldsymbol{x}_0) > 0$. Hence

$$(P\boldsymbol{x}, \boldsymbol{x}) \geq q_0 := q(\boldsymbol{x}_0), \quad \forall \boldsymbol{x} \in M.$$

Equivalently, this means that

$$\left(P\frac{1}{\|\boldsymbol{x}\|}\boldsymbol{x}, \frac{1}{\|\boldsymbol{x}\|}\boldsymbol{x} \right) \geq q_0, \quad \forall \boldsymbol{x} \in \mathbb{R}^n \setminus \{0\}. \tag{A.16}$$

The last inequality combined with the properties of the scalar product implies (A.15) with $\omega = q_0$.

Lemma A.4 [The Cauchy–Schwartz inequality] *If P is a real, $n \times n$ symmetric and positive matrix, then*

$$\left| (P\boldsymbol{x}, \boldsymbol{y}) \right| \leq (P\boldsymbol{x}, \boldsymbol{x})^{\frac{1}{2}} (P\boldsymbol{y}, \boldsymbol{y})^{\frac{1}{2}}, \quad \forall \boldsymbol{x}, \boldsymbol{y} \in \mathbb{R}^n. \tag{A.17}$$

In particular, for $P = \mathbb{1}_e$, we have

$$\left| (\boldsymbol{x}, \boldsymbol{y}) \right| \leq \|\boldsymbol{x}\|_e \, \|\boldsymbol{y}\|_e, \quad \forall \boldsymbol{x}, \boldsymbol{y} \in \mathbb{R}^n. \tag{A.18}$$

Proof Consider the real-valued function

$$\psi(\lambda) = \big(P(\boldsymbol{x}+\lambda \boldsymbol{y}), \boldsymbol{x}+\lambda \boldsymbol{y}, \big), \quad \lambda \in \mathbb{R}.$$

From the symmetry of P and properties of the scalar product, we deduce that for any $\boldsymbol{x}, \boldsymbol{y} \in \mathbb{R}^n$ the function $\psi(\lambda)$ is the quadratic polynomial

$$\psi(\lambda) = \lambda(P\boldsymbol{y}, \boldsymbol{y}) + 2\lambda(P\boldsymbol{x}, \boldsymbol{y}) + (P\boldsymbol{x}, \boldsymbol{x}), \quad \lambda \in \mathbb{R}.$$

Since $\psi(\lambda) \geq 0, \forall \lambda \in \mathbb{R}$, it follows that the discriminant of the quadratic polynomial ψ is non-positive. This is precisely the content of (A.17).

Lemma A.5 *Let P be a real, symmetric, positive $n \times n$ matrix. There exists a $C > 0$ such that*

$$\|P\boldsymbol{x}\|_{\mathbf{e}}^2 \leq C(P\boldsymbol{x}, \boldsymbol{x})^{\frac{1}{2}} \|\boldsymbol{x}\|_{\mathbf{e}}, \quad \forall \boldsymbol{x} \in \mathbb{R}^n. \tag{A.19}$$

Proof From Lemma A.4, we deduce that

$$\|P\boldsymbol{x}\|_{\mathbf{e}}^2 = (P\boldsymbol{x}, P\boldsymbol{x}) \leq (P\boldsymbol{x}, \boldsymbol{x})^{\frac{1}{2}} (P^2\boldsymbol{x}, P\boldsymbol{x})^{\frac{1}{2}} \leq K\|P\|^{\frac{3}{2}} \|\boldsymbol{x}\|_{\mathbf{e}} (P\boldsymbol{x}, \boldsymbol{x})^{\frac{1}{2}}.$$

Lemma A.6 *Let $\boldsymbol{x} : [a, b] \to \mathbb{R}^1$ be a C^1-function and P a real, symmetric $n \times n$ matrix. Then*

$$\frac{1}{2}\frac{d}{dt}\big(P\boldsymbol{x}(t)\boldsymbol{x}(t) \big) = \big(P\boldsymbol{x}(t), \boldsymbol{x}'(t) \big), \quad \forall t \in [a, b]. \tag{A.20}$$

Proof By definition

$$\begin{aligned}
\frac{d}{dt}(P\boldsymbol{x}(t), \boldsymbol{x}(t)) &= \lim_{h\to 0}\frac{1}{h}[(P\boldsymbol{x}(t+h) - x(t), \boldsymbol{x}(t+h) - (P\boldsymbol{x}(t), \boldsymbol{x}(t)))] \\
&= \lim_{h\to 0}\left(P\frac{1}{h}(\boldsymbol{x}(t+h) - x(t)), \boldsymbol{x}(t+h) \right) + \left(P\boldsymbol{x}(t), \frac{1}{h}(\boldsymbol{x}(t+h) - x(t)) \right) \\
&= (P\boldsymbol{x}'(t), \boldsymbol{x}(t)) + (P\boldsymbol{x}(t), \boldsymbol{x}'(t)) = 2(P\boldsymbol{x}(t), \boldsymbol{x}'(t)), \ \forall t \in [a, b].
\end{aligned}$$

A.3 The Arzelà Theorem

Let $I = [a, b]$ be a compact interval of the real axis. We will denote by $C(I; \mathbb{R}^n)$ or $C([a, b]; \mathbb{R}^n)$ the space of continuous functions $I \to \mathbb{R}^n$.

The space $C(I; \mathbb{R}^n)$ has a natural vector space structure with respect to the natural operations on functions. It is equipped with the *uniform norm*

$$\|\boldsymbol{x}\|_{\mathbf{u}} := \sup_{t\in I} \|\boldsymbol{x}(t)\|. \tag{A.21}$$

It is not hard to see that the convergence in norm (A.21) of a sequence of functions $\{\boldsymbol{x}_j\} \subset C(I;\mathbb{R}^n)$ is equivalent to the uniform convergence of this sequence on the interval I.

Definition A.2 A set $\mathcal{M} \subset C(I;\mathbb{R}^n)$ is called *bounded* if there exists a constant $M > 0$ such that

$$\|\boldsymbol{x}\|_{\mathbf{u}} \leq M, \quad \forall \boldsymbol{x} \in \mathcal{M}. \tag{A.22}$$

The set $\mathcal{M}$ is called *uniformly equicontinuous* if

$$\begin{aligned} \forall \varepsilon > 0,\ \exists \delta = \delta(\varepsilon) > 0 : \|\boldsymbol{x}(t) - \boldsymbol{x}(s)\| \leq \varepsilon,\ \forall \boldsymbol{x} \in \mathcal{M}, \\ \forall t, s \in I,\ |t - s| \leq \delta. \end{aligned} \tag{A.23}$$

Our next theorem, due to *C. Arzelà* (1847–1912), is a compactness result in the space $C(I, \mathbb{R}^n)$ similar to the well-known Bolzano–Weierstrass theorem.

Theorem A.1 *Suppose that $\mathcal{M} \subset C(I, \mathbb{R}^n)$ is bounded and uniformly equicontinuous. Then, any sequence of functions in $\mathcal{M}$ contains a subsequence that is uniformly convergent on I.*

Proof The set $\mathbb{Q} \cap I$ of rational numbers in the interval I is countable and thus we can describe it as consisting of the terms of a sequence $(r_k)_{k\geq 1}$. Consider the set

$$M_1 := \big\{\boldsymbol{x}(r_1);\ \ \boldsymbol{x} \in \mathcal{M}\big\} \subset \mathbb{R}.$$

The set M_1 is obviously bounded and thus, according to the Bolzano–Weierstrass theorem, it admits a bounded subsequence

$$\boldsymbol{x}_1^2(r_1), \boldsymbol{x}_1^2(r_1), \ldots, \boldsymbol{x}_1^m(r_1), \ldots.$$

Next, consider the set

$$M_2 := \big\{\boldsymbol{x}_1^2(r_2), \boldsymbol{x}_1^2(r_2), \ldots, \boldsymbol{x}_1^m(r_2), \ldots\big\} \subset \mathbb{R}.$$

It is bounded, and we deduce again that it contains a convergent subsequence $\{x_2^j(r_2)\}_{j>1}$. Iterating this procedure, we obtain an infinite array

$$\begin{array}{llll} \boldsymbol{x}_1^1 & \boldsymbol{x}_1^2 & \cdots \boldsymbol{x}_1^m & \cdots \\ \boldsymbol{x}_2^1 & \boldsymbol{x}_2^2 & \cdots \boldsymbol{x}_2^m & \cdots \\ \cdots & \cdots & \cdots\cdots & \cdots \\ \boldsymbol{x}_m^1 & \boldsymbol{x}_m^2 & \cdots \boldsymbol{x}_m^m & \cdots \end{array} \tag{A.24}$$

Every row of this array is a subsequence of the row immediately above it and the sequence of functions on the m-th row converges on the finite set $\{r_1, \ldots, r_m\}$. We deduce that the diagonal sequence $\{\boldsymbol{x}_m^m\}_{m\geq 1}$ converges on $\mathbb{Q} \cap I$. We will prove that this sequence converges uniformly on I.

The uniform equicontinuity condition shows that, for any $\varepsilon > 0$, there exists a $\delta(\varepsilon) > 0$ such that

$$\|\boldsymbol{x}_m^m(t) - \boldsymbol{x}_m^m(s)\| \leq \varepsilon, \quad \forall |t-s| \leq \delta(\varepsilon), \quad \forall m. \tag{A.25}$$

Since $\mathbb{Q}$ is dense in $\mathbb{R}$, we deduce that there exists an $N = N(\varepsilon)$ with the property that any $t \in I$ is within $\delta(\varepsilon)$ from at least one of the points $r_1, \dots, r_{N(\varepsilon)}$,

$$\min_{1 \leq i \leq N(\varepsilon)} |t - r_i| \leq \delta(\varepsilon).$$

Inequalities (A.25) show that, for arbitrary k, ℓ and any $t \in I$, we have

$$\begin{aligned} &\|\boldsymbol{x}_k^k(t) - \boldsymbol{x}_\ell^\ell(t)\| \\ &\leq \min_{1 \leq i \leq N(\varepsilon)} \left(\|\boldsymbol{x}_k^k(t) - \boldsymbol{x}_k^k(r_i)\| + \|\boldsymbol{x}_k^k(r_i) - \boldsymbol{x}_\ell^\ell(r_i)\| + \|\boldsymbol{x}_\ell^\ell(r_i) - \boldsymbol{x}_\ell^\ell(t)\|\right) \\ &\leq 2\varepsilon + \max_{1 \leq i \leq N(\varepsilon)} \|\boldsymbol{x}_k^k(r_i) - \boldsymbol{x}_\ell^\ell(r_i)\|. \end{aligned}$$

The sequence $(x_k^k(r_i))_{k \geq 1}$ converges for any $1 \leq i \leq N(\varepsilon)$. It is thus Cauchy for each such i. In particular, we can find $K = K(\varepsilon) > 0$ such that

$$\max_{1 \leq i \leq N(\varepsilon)} \|\boldsymbol{x}_k^k(r_i) - \boldsymbol{x}_\ell^\ell(r_i)\| \leq \varepsilon, \quad \forall k, \ell \geq K(\varepsilon).$$

Hence

$$\|\boldsymbol{x}_k^k(t) - \boldsymbol{x}_\ell^\ell(t)\| \leq 3\varepsilon, \quad \forall t \in I, \quad k, \ell \geq K(\varepsilon). \tag{A.26}$$

The above inequalities show that the sequence of functions $(\boldsymbol{x}_k^k)_{k \geq 1}$ satisfies the conditions in Cauchy's criterion of uniform convergence, and thus it is uniformly convergent on I.

Remark A.1 Consider a set $\mathcal{M} \subset C(I; \mathbb{R}^n)$ consisting of differentiable functions such that there exists a $C > 0$ with the property

$$\|\boldsymbol{x}'(t)\| \leq C, \quad \forall t \in I, \quad \forall \boldsymbol{x} \in \mathcal{M}.$$

Then $\mathcal{M}$ is uniformly equicontinuous. Indeed, for any $\boldsymbol{x} \in \mathcal{M}$ and any $t, s \in I$, $s < t$,

$$\|\boldsymbol{x}(t) - \boldsymbol{x}(s)\| \leq \int_s^t \|\boldsymbol{x}'(\tau)\| d\tau \leq C|t-s|.$$

A.1 Does the family of functions $\{\sin nt; \quad n = 1, 2, \dots\}$ satisfy the assumptions of Arzelà's theorem on $[0, \pi]$?

A.4 The Contraction Principle

Suppose that X is a set and $d : X \times X \to [0, \infty)$ is a nonnegative function on the Cartesian product $X \times X$. We say that d defines a *metric* on X if it satisfies the following conditions.

$$d(x, y) \geq 0, \quad \forall x, y \in X, \tag{A.27}$$

$$d(x, y) = 0 \iff x = y, \tag{A.28}$$

$$d(x, y) = d(y, x), \quad \forall x, y \in X, \tag{A.29}$$

$$d(x, z) \leq d(x, y) + d(y, z), \quad \forall x, y, z \in X. \tag{A.30}$$

A set equipped with a metric is called a *metric space*.

A metric space (X, d) is equipped with a natural topology. Indeed, every point $x_0 \in X$ admits a system of neighborhoods consisting of sets of the form $S(x_0, r)$, where $S(x_0, r)$ is the open ball of radius r centered at x_0, that is,

$$S(x_0, r) := \big\{ x \in X; \quad d(x, x_0) < r \big\}. \tag{A.31}$$

In particular, we can define the concept of *convergence*. We say that the sequence $\{x_n\}_{n \geq 1} \subset X$ converges to $x \in X$ as $n \to \infty$ if

$$\lim_{n \to \infty} d(x_n, x) = 0.$$

The sequence $\{x_n\}_{n \geq 1}$ is called *fundamental* if for any $\varepsilon > 0$ there exists an $N(\varepsilon) > 0$ such that

$$d(x_n, x_m) \leq \varepsilon, \quad \forall m, n \geq N(\varepsilon).$$

It is not hard to see that any convergent sequence is fundamental. The converse is not necessarily true.

Definition A.3 A metric space (X, d) is called *complete* if any fundamental sequence in X is convergent.

Example A.1 (i) The space $X = \mathbb{R}^n$ with the metric $d(\boldsymbol{x}, \boldsymbol{y}) = \|\boldsymbol{x} - \boldsymbol{y}\|$ is complete.
(ii) For any real number α, the space $X = C([a, b]; \mathbb{R}^n)$ equipped with the metric

$$d(\boldsymbol{x}, \boldsymbol{y}) = \sup_{t \in [a,b]} \|\boldsymbol{x}(t) - \boldsymbol{y}(t)\| e^{\alpha t}, \quad \boldsymbol{x}, \boldsymbol{y} \in X, \tag{A.32}$$

is complete. Let us observe that convergence in the metric (A.32) is equivalent to uniform convergence on the compact interval $[a, b]$.

Definition A.4 Let (X, d) be a metric space. A mapping $\Gamma : X \to X$ is called a *contraction* if there exists a $\rho \in (0, 1)$ such that

$$d(\Gamma x, \Gamma y) \leq \rho d(x, y), \quad \forall x, y \in X.$$

An element $x_0 \in X$ is called a *fixed point* of Γ if

$$\Gamma x_0 = x_0.$$

The next theorem is known as the *contraction principle* or *Banach's fixed point theorem* (*S. Banach* (1892–1945)).

Theorem A.2 *If* (X, d) *is a* complete *metric space and* $\Gamma : X \to X$ *is a* contraction *on* X*, then* Γ *admits a unique fixed point.*

Proof Fix $x_1 \in X$ and consider the sequence of successive approximations

$$x_{n+1} = \Gamma x_n, \quad n = 1, 2, \dots . \tag{A.33}$$

Since Γ is a contraction, we deduce that

$$d(x_{n+1}, x_n) \leq \rho d(\Gamma x_n, \Gamma x_{n-1}) \leq \cdots \leq \rho^{n-1} d(x_2, x_1).$$

Using the triangle inequality (A.30) iteratively, we deduce that

$$d(x_{n+p}, x_n) \leq \sum_{j=n}^{n+p-1} d(x_{j+1}, x_j) \leq \left(\sum_{j=n}^{n+p-1} \rho^{j-1} \right) d(x_2, x_1), \tag{A.34}$$

for any positive integers n, p. Since $\rho \in (0, 1)$, we deduce that the geometric series

$$\sum_{j=0}^{\infty} \rho^j$$

is convergent, and (A.34) implies that the sequence $\{x_n\}_{n \geq 1}$ is fundamental. The space X is complete and thus this sequence converges as $n \to \infty$ to some point $x_\infty \in X$. Since Γ is a contraction, we deduce that

$$d(\Gamma x_n, \Gamma x_\infty) \leq \rho d(x_n, x_\infty), \quad \forall n \geq 1.$$

Letting $n \to \infty$ in the above inequality, we have

$$\lim_{n \to \infty} \Gamma x_n = \Gamma x_\infty.$$

Letting $n \to \infty$ in (A.33), we deduce that $x_\infty = \Gamma_\infty$. Thus x_∞ is a fixed point of Γ.

The uniqueness of the fixed point follows from the contraction property. Indeed, if x_0, y_0 are two fixed points, then

$$d(x_0, y_0) = d(\Gamma x_0, \Gamma y_0) \leq \rho d(x_0, y_0).$$

Since $\rho \in (0, 1)$, we conclude that $d(x_0, y_0) = 0$, and thus $x_0 = y_0$.

Example A.2 As an application of the contraction principle, we present an alternative proof of the existence and uniqueness result in Theorem 2.4. We will make the same assumptions and we will use the same notations as in Theorem 2.4.

We consider the set

$$X := \{x \in C(T; \mathbb{R}^n);\ \ \|x(t) - x_0\| \leq b\},\ \ I = [t_0 - \delta, t_0 + \delta], \tag{A.35}$$

equipped with the metric

$$d(x, y) = \sup_{t \in I} \|x(t) - y(t)\| e^{-2Lt}, \tag{A.36}$$

where L is the Lipschitz constant of f. On X, we define the operator

$$(\Gamma x)(t) := x_0 + \int_{t_0}^{t} f(s, x(s))ds,\ \ t \in I. \tag{A.37}$$

Observe that (2.15) implies that

$$\|\Gamma x(t) - x_0\| \leq b,\ \ \forall t \in I,$$

that is, Γ maps the space X to itself. On the other hand, any elementary computation based on the inequality (2.14) leads to

$$\|(\Gamma x)(t) - (\Gamma y)(t)\| \leq L \left| \int_{t_0}^{t} \|x(s) - y(s)\| ds \right|,$$

and, using (A.36), we deduce

$$d(\Gamma x, \Gamma y) \leq \frac{1}{2} d(x, y),\ \ \forall x, y, \in X.$$

From Theorem A.2, it follows that there exists a unique $x \in X$ such that $\Gamma x = x$. In other words, the integral equation

$$x(t) = x_0 + \int_{t_0}^{t} f(s, x(s))ds,\ \ t \in I,$$

has a unique solution in X.

A.5 Differentiable Functions and the Implicit Function Theorem

Let $\boldsymbol{f} : \Omega \to \mathbb{R}^m$ be a function defined on an open subset $\Omega \subset \mathbb{R}^n$ and valued in $\mathbb{R}^m$. There will be no danger of confusion if we denote by the same symbol the norms in $\mathbb{R}^n$ and $\mathbb{R}^m$.

If $\boldsymbol{x}_0 \in \Omega$, we say that

$$\lim_{\boldsymbol{x}\to\boldsymbol{x}_0} \boldsymbol{f}(\boldsymbol{x}) = \boldsymbol{u}$$

if

$$\lim_{\boldsymbol{x}\to\boldsymbol{x}_0} \|\boldsymbol{f}(\boldsymbol{x}) - \boldsymbol{u}\| = 0.$$

The function $\boldsymbol{f}$ is called *continuous* at $\boldsymbol{x}_0$ if

$$\lim_{\boldsymbol{x}\to\boldsymbol{x}_0} \boldsymbol{f}(\boldsymbol{x}) = \boldsymbol{f}(\boldsymbol{x}_0).$$

The function $\boldsymbol{f}$ is called differentiable at $\boldsymbol{x}_0$ if there exists a linear map $\mathbb{R}^n \to \mathbb{R}^m$, denoted by

$$\boldsymbol{f}'(\boldsymbol{x}_0), \quad \boldsymbol{f}_{\boldsymbol{x}}(\boldsymbol{x}_0), \quad \text{or} \quad \frac{\partial \boldsymbol{f}}{\partial \boldsymbol{x}}(\boldsymbol{x}_0),$$

and called the *derivative* of $\boldsymbol{f}$ at $\boldsymbol{x}_0$, such that

$$\lim_{\boldsymbol{x}\to\boldsymbol{x}_0} \frac{\|\boldsymbol{f}(\boldsymbol{x}) - \boldsymbol{f}(\boldsymbol{x}_0) - \boldsymbol{f}'(\boldsymbol{x}_0)(\boldsymbol{x} - \boldsymbol{x}_0)\|}{\|\boldsymbol{x} - \boldsymbol{x}_0\|} = 0. \tag{A.38}$$

Lemma A.1 shows that the above definition is independent of the norms on the spaces $\mathbb{R}^n$ and $\mathbb{R}^m$. When $m = 1$, so that $\boldsymbol{f}$ is scalar-valued, we set

$$\text{grad } \boldsymbol{f} := \boldsymbol{f}',$$

and we will refer to this derivative as the *gradient* of $\boldsymbol{f}$.

The function $\boldsymbol{f} : \Omega \to \mathbb{R}^m$ is said to be *continuously differentiable*, or C^1, if it is differentiable at every point $\boldsymbol{x} \in \Omega$, and then the resulting map

$$\boldsymbol{x} \mapsto \boldsymbol{f}'_{\boldsymbol{x}} \in \text{Hom}(\mathbb{R}^n, \mathbb{R}^m) := \text{the vector space of linear operators } \mathbb{R}^n \to \mathbb{R}^m$$

is continuous. More precisely, if $\boldsymbol{f}$ is represented as a column vector

$$\boldsymbol{f}(\boldsymbol{x}) = \begin{bmatrix} f_1(\boldsymbol{x}) \\ \vdots \\ f_m(\boldsymbol{x}) \end{bmatrix},$$

then the continuity and differentiability of $\boldsymbol{f}$ is, respectively, equivalent to the continuity and differentiability of each of the components $f_i(\boldsymbol{x}), i = 1, \ldots, m$. Moreover, the derivative $\boldsymbol{f}'_{\boldsymbol{x}}$ is none other than the Jacobian matrix of the system of functions $f_1, \ldots, f_m$, that is,

$$\boldsymbol{f}'(\boldsymbol{x}) = \left[\frac{\partial f_i}{\partial x_j}(\boldsymbol{x}) \right]_{\substack{1 \le i \le m \\ 1 \le j \le n}}. \tag{A.39}$$

Theorem A.3 (The implicit function theorem) *Suppose that $U \subset \mathbb{R}^m$ and $V \subset \mathbb{R}^n$ are two open sets and $F : U \times V \to \mathbb{R}^m$ is a C^1-mapping. Assume, additionally, that there exists a point $(\boldsymbol{x}_0, \boldsymbol{y}_0) \in U \times V$ such that*

$$F(\boldsymbol{x}_0, \boldsymbol{y}_0) = 0, \quad \det F_{\boldsymbol{x}}(\boldsymbol{x}_0, \boldsymbol{y}_0) \neq 0. \tag{A.40}$$

Then there exists an open neighborhood Ω of $\boldsymbol{y}_0$ in V, an open neighborhood $\mathcal{O}$ of $\boldsymbol{x}_0 \in U$, and a continuous function $f : \Omega \to \mathcal{O}$ such that $\boldsymbol{f}(\boldsymbol{y}_0) = \boldsymbol{x}_0$ and

$$F(\boldsymbol{x}, \boldsymbol{y}) = 0, \quad (\boldsymbol{x}, \boldsymbol{y}) \in \mathcal{O} \times \Omega \Longleftrightarrow \boldsymbol{x} = f(\boldsymbol{y}). \tag{A.41}$$

Moreover,

$$\det F_{\boldsymbol{x}}\big(f(\boldsymbol{y}), \boldsymbol{y}\big) \neq 0, \quad \forall \boldsymbol{y} \in \Omega,$$

the function f is C^1 on Ω, and

$$\boldsymbol{f}'(\boldsymbol{y}) = -F_{\boldsymbol{x}}\big(\boldsymbol{f}(\boldsymbol{y}), \boldsymbol{y}\big)^{-1} F_{\boldsymbol{y}}\big(\boldsymbol{f}(\boldsymbol{y}), \boldsymbol{y}\big), \quad \forall \boldsymbol{y} \in \Omega. \tag{A.42}$$

Proof We denote by $\| - \|_m$ and $\| - \|_n$ two norms in $\mathbb{R}^m$ and $\mathbb{R}^n$ respectively. Without loss of generality, we can assume that

- $\boldsymbol{x}_0 = 0$ and $\boldsymbol{y}_0 = 0$.
- The set U is an open ball of radius $r_U < 1$ in $\mathbb{R}^m$ centered at $\boldsymbol{x}_0$, and the set V is an open ball of radius $r_V < 1$ in $\mathbb{R}^n$ centered at $\boldsymbol{y}_0$.
- For any $(\boldsymbol{x}, \boldsymbol{y}) \in U \times V$, the partial derivative $F_{\boldsymbol{x}}(\boldsymbol{x}, \boldsymbol{y})$ is invertible.

We have the equality

$$F(\boldsymbol{x}, \boldsymbol{y}) = F_x(0, 0)\boldsymbol{x} + F_{\boldsymbol{y}}(0, 0)\boldsymbol{y} + R(\boldsymbol{x}, \boldsymbol{y}) + R(\boldsymbol{x}, \boldsymbol{y}), \quad \forall (\boldsymbol{x}, \boldsymbol{y}) \in U \times V, \tag{A.43}$$

where $F_{\boldsymbol{x}}$ and $F_{\boldsymbol{y}}$ are the partial derivatives of F with respect to $\boldsymbol{x}$ and $\boldsymbol{y}$ respectively. The function $R(\boldsymbol{x}, \boldsymbol{y})$ is obviously C^1 on $U \times V$ and, additionally, we have

$$R(0, 0) = 0, \quad R_{\boldsymbol{x}}(0, 0) = 0, \quad R_{\boldsymbol{y}}(0, 0) = 0. \tag{A.44}$$

Since $R' = (R_{\boldsymbol{x}}, R_{\boldsymbol{y}})$ is continuous on $U \times V$, we deduce that for any $\varepsilon > 0$ there exists a $\delta(\varepsilon) > 0$ such that

$$\| R'(\boldsymbol{x}, \boldsymbol{y}) \| \leq \varepsilon, \quad \forall \|\boldsymbol{x}\|_m + \|\boldsymbol{y}\|_n \leq \delta(\varepsilon).$$

Taking into account definition (A.38) of the derivative R', we deduce that, for any $(x_i, y_i) \in U \times V, i = 1, 2$, such that

$$\|\boldsymbol{x}_i\|_m + \|\boldsymbol{y}_i\|_n \leq \delta(\varepsilon), \quad i = 1, 2,$$

we have

$$\|R(\boldsymbol{x}_1, \boldsymbol{y}_1) - R(\boldsymbol{x}_2, \boldsymbol{y}_2)\|_m \leq \varepsilon\big(\|\boldsymbol{x}_1 - \boldsymbol{x}_2\|_m + \|\boldsymbol{y}_1 - \boldsymbol{y}_2\|_n \big). \tag{A.45}$$

Let $G : U \times V \to \mathbb{R}^m$ be the function

$$G(\boldsymbol{x}, \boldsymbol{y}) = -AF_{\boldsymbol{y}}(0, 0)\boldsymbol{y} - AR(\boldsymbol{x}, \boldsymbol{y}), \quad A := F_{\boldsymbol{x}}(0, 0)^{-1}. \tag{A.46}$$

The equation $F(\boldsymbol{x}, \boldsymbol{y}) = 0$ we are interested in can be rewritten as a fixed point problem

$$\boldsymbol{x} = -G(\boldsymbol{x}, \boldsymbol{y}).$$

From (4.77) and the equality $G(0, 0) = 0$, we deduce that

$$\|G(\boldsymbol{x}, \boldsymbol{y})\|_m \leq \|A\| \, \|F_{\boldsymbol{y}}(0, 0)\| \, \|\boldsymbol{y}\|_n + \varepsilon\|A\|\big(\|\boldsymbol{x}\|_m + \|\boldsymbol{y}\|_n \big) \tag{A.47}$$

and

$$\|G(\boldsymbol{x}_1, \boldsymbol{y}) - G(\boldsymbol{x}_2, \boldsymbol{y})\|_m \leq \varepsilon\|A\| \, \|\boldsymbol{x}_1 - \boldsymbol{x}_2\|_m, \tag{A.48}$$

for $\|\boldsymbol{x}\|_m + \|\boldsymbol{y}\|_n \leq \delta(\varepsilon)$, $\|\boldsymbol{x}_i\|_m + \|\boldsymbol{y}\|_n \leq \delta(\varepsilon)$, $i = 1, 2$.

We set

$$\varepsilon := \frac{1}{2} \min\big(r_U, \|A\|^{-1}\big),$$
$$\eta := \frac{\min(\delta(\varepsilon), r_V)}{2\big(\|F_{\boldsymbol{y}}(0, 0)\|\varepsilon^{-1} + 1\big)},$$

and we consider the open balls

$$\mathcal{O} := \big\{ \boldsymbol{x} \in U; \ \|\boldsymbol{x}\|_m < \delta(\varepsilon) \big\}, \tag{A.49}$$
$$\Omega := \big\{ \boldsymbol{y} \in V; \ \|\boldsymbol{y}\|_n < \eta \big\}. \tag{A.50}$$

We denote their closures by $\overline{\mathcal{O}}$ and $\overline{\Omega}$ respectively. From (A.47) and (A.48), we deduce that for any $\boldsymbol{y} \in \overline{\Omega}$ we have

$$G(\boldsymbol{x}, \boldsymbol{y}) \in \overline{\mathcal{O}}, \quad \forall \boldsymbol{x} \in \overline{\mathcal{O}}, \tag{A.51}$$
$$\|G(\boldsymbol{x}_1, \boldsymbol{y}) - G(\boldsymbol{x}_2, \boldsymbol{y})\|_m \leq \frac{1}{2}\|\boldsymbol{x}_1 - \boldsymbol{x}_1\|_m, \quad \forall \boldsymbol{x}_1, \boldsymbol{x}_2 \in \overline{\mathcal{O}}. \tag{A.52}$$

This shows that for any $\boldsymbol{y} \in \overline{\Omega}$ we have a contraction

$$T_{\boldsymbol{y}} : \overline{\mathcal{O}} \to \overline{\mathcal{O}}, \quad \boldsymbol{x} \mapsto T_{\boldsymbol{y}}\boldsymbol{x} := G(\boldsymbol{x}, \boldsymbol{y}).$$

This has a unique fixed point that we denote by $\boldsymbol{f}(\boldsymbol{y})$. Note that a point $(\boldsymbol{x}, \boldsymbol{y}) \in \overline{\mathcal{O}} \times \overline{\Omega}$ is a solution of the equation $F(\boldsymbol{x}, \boldsymbol{y}) = 0$ if and only if $\boldsymbol{x} = \boldsymbol{f}(\boldsymbol{y})$.

Let us observe that the map $\boldsymbol{f} : \overline{\Omega} \to \overline{\mathcal{O}}$ is Lipschitz continuous. Indeed, if $\boldsymbol{y}_1, \boldsymbol{y}_2 \in \overline{\Omega}$, then

$$\begin{aligned}
&\|\boldsymbol{f}(\boldsymbol{y}_1) - \boldsymbol{f}(\boldsymbol{y}_2)\|_m = \big\| G(\boldsymbol{f}(\boldsymbol{y}_1), \boldsymbol{y}_1) - G(\boldsymbol{f}(\boldsymbol{y}_2), \boldsymbol{y}_2) \big\|_m \\
&\leq \|AF_{\boldsymbol{y}}(0,0)\boldsymbol{y}_1 - AF_{\boldsymbol{y}}(0,0)\boldsymbol{y}_2\| + \|A\| \, \|R(\boldsymbol{f}(\boldsymbol{y}_1), \boldsymbol{y}_1) - R(\boldsymbol{f}(\boldsymbol{y}_2), \boldsymbol{y}_2)\|_m \\
&\overset{(A.45)}{\leq} \|AF_{\boldsymbol{y}}(0,0)\| \, \|\boldsymbol{y}_1 - \boldsymbol{y}_1\|_n + \varepsilon\|A\| \Big(\|\boldsymbol{f}(\boldsymbol{y}_1) - \boldsymbol{f}(\boldsymbol{y}_2)\|_m + |\boldsymbol{y}_1 - \boldsymbol{y}_2\|_n \Big) \\
&\leq \|A\| \big(\varepsilon + \|F_{\boldsymbol{y}}(0,0)\| \big) \|\boldsymbol{y}_1 - \boldsymbol{y}_1\|_n + \frac{1}{2} \|\boldsymbol{f}(\boldsymbol{y}_1) - \boldsymbol{f}(\boldsymbol{y}_2)\|_m,
\end{aligned}$$

so that

$$\|\boldsymbol{f}(\boldsymbol{y}_1) - \boldsymbol{f}(\boldsymbol{y}_2)\|_m \leq 2\|A\| \big(\varepsilon + \|F_{\boldsymbol{y}}(0,0)\| \big) \, \|\boldsymbol{y}_1 - \boldsymbol{y}_1\|_n.$$

To prove that the function $\boldsymbol{f}$ is differentiable on Ω, we use the defining property of $\boldsymbol{f}(\boldsymbol{y})$,

$$F(\boldsymbol{f}(\boldsymbol{y}), \boldsymbol{y}) = 0.$$

Since $\boldsymbol{f}$ is Lipschitz continuous, we deduce that for $\boldsymbol{y}_0 \in \Omega$ we have

$$\begin{aligned}
0 &= F\big(\boldsymbol{f}(\boldsymbol{y}_0 + \boldsymbol{h}), \boldsymbol{y}_0 + \boldsymbol{h}\big) - F\big(\boldsymbol{f}(\boldsymbol{y}_0), \boldsymbol{y}_0\big) \\
&= F_{\boldsymbol{x}}\big(\boldsymbol{f}(\boldsymbol{y}_0), \boldsymbol{y}_0\big)\big(\boldsymbol{f}(\boldsymbol{y}_0 + \boldsymbol{h}) - \boldsymbol{f}(\boldsymbol{y}_0)\big) + F_{\boldsymbol{y}}\big(\boldsymbol{f}(\boldsymbol{y}_0), \boldsymbol{y}_0\big)\boldsymbol{h} + o\big(\|\boldsymbol{h}\|_n\big).
\end{aligned}$$

Hence,

$$\begin{aligned}
&\boldsymbol{f}(\boldsymbol{y}_0 + \boldsymbol{h}) - \boldsymbol{f}(\boldsymbol{y}_0) \\
&\quad = -F_{\boldsymbol{x}}(\boldsymbol{f}(\boldsymbol{y}_0), \boldsymbol{y}_0)^{-1} F_{\boldsymbol{y}}(\boldsymbol{f}(\boldsymbol{y}_0), \boldsymbol{y}_0)\boldsymbol{h} + F_{\boldsymbol{x}}(\boldsymbol{f}(\boldsymbol{y}_0), \boldsymbol{y}_0)^{-1} o(\|\boldsymbol{h}\|_n).
\end{aligned}$$

This shows that $\boldsymbol{f}$ is differentiable at $\boldsymbol{y}_0$ and its derivative is given by (A.42).

Corollary A.1 *Let $\boldsymbol{g} : U \to \mathbb{R}^m$ be a C^1-function defined on the open subset $U \subset \mathbb{R}^m$. We assume that there exists an $\boldsymbol{x}_0 \in U$ such that*

$$\det \boldsymbol{g}'_{\boldsymbol{x}}(\boldsymbol{x}_0) \neq 0. \tag{A.53}$$

Then there exists an open neighborhood $\mathcal{O}$ of $\boldsymbol{x}_0 \in U$ such that $\boldsymbol{g}$ maps $\mathcal{O}$ bijectively onto a neighborhood Ω of $\boldsymbol{g}(\boldsymbol{x}_0)$. Moreover, the inverse map $\boldsymbol{g}^{-1} : \Omega \to \mathcal{O}$ is also C^1. In other words, $\boldsymbol{g}$ induces a C^1-diffeomorphism $\mathcal{O} \to \Omega$.

Proof Apply the implicit function theorem to the function

$$F(\boldsymbol{x}, \boldsymbol{y}) = \boldsymbol{y} - \boldsymbol{g}(\boldsymbol{x}).$$

Remark A.2 In more concrete terms, Theorem A.3 states that the solutions of the (underdetermined) system

$$\begin{aligned} F_1(x_1, \ldots, x_m; y_1, \ldots, y_n) &= 0 \\ &\vdots \\ F_m(x_1, \ldots, x_m; y_1, \ldots, y_n) &= 0 \end{aligned} \tag{A.54}$$

form a family with n independent parameters $y_1, \ldots, y_n$ while the remaining unknown $x_1, \ldots, x_m$ can be described as differentiable functions of the parameters $\boldsymbol{y}$,

$$x_i = f_i(y_1, \ldots, y_n), \quad i = 1, \ldots, m, \tag{A.55}$$

in a neighborhood of a point

$$(\boldsymbol{x}^*, \boldsymbol{y}^*) = (x_1^*, \ldots, x_m^*; y_1^*, \ldots, y_n^*)$$

such that

$$F_i(\boldsymbol{x}^*, \boldsymbol{y}^*) = 0, \quad \forall i = 1, \ldots, m, \tag{A.56}$$

$$\det \frac{D(F_1, \ldots, F_m)}{D(x_1, \ldots, x_m)}(\boldsymbol{x}^*, \boldsymbol{y}^*) \neq 0. \tag{A.57}$$

To use the classical terminology, we say that the functions $f_1, \ldots, f_m$ are *implicitly* defined by system (A.54).

A.6 Convex Sets and Functions

A subset K of the space $\mathbb{R}^n$ is called *convex* if, for any two points $\boldsymbol{x}, \boldsymbol{y} \in K$, the line segment they determine is also contained in K. In other words, if $\boldsymbol{x}, \boldsymbol{y} \in K$, then

$$t\boldsymbol{x} + (1 - t)\boldsymbol{y} \in K, \quad \forall t \in [0, 1]. \tag{A.58}$$

A function $f : \mathbb{R}^n \to (-\infty, \infty]$ is called *convex* if

$$f\big(t\boldsymbol{x} + (1 - t)\boldsymbol{y}\big) \leq tf(\boldsymbol{x}) + (1 - t)f(\boldsymbol{y}), \quad \forall t \in [0, 1], \quad \boldsymbol{x}, \boldsymbol{y} \in \mathbb{R}^n. \tag{A.59}$$

The *Legendre transform* associates to a convex function $f : \mathbb{R}^n \to (-\infty, \infty]$ the function $f^* : \mathbb{R}^n \to (-\infty, \infty]$ given by

$$f^*(\boldsymbol{p}) = \sup\big\{ (\boldsymbol{p}, \boldsymbol{x}) - f(\boldsymbol{x});\ \ \boldsymbol{x} \in \mathbb{R}^n \big\}. \tag{A.60}$$

The function f^* is called the *conjugate* of f. In (A.60), $(-, -)$ denotes the canonical scalar product on $\mathbb{R}^n$.

We mention without proof the following result.

Proposition A.1 *A convex function $f : \mathbb{R}^n \to (-\infty, \infty]$ that is everywhere finite is continuous.*

Let us assume that the function $f : \mathbb{R}^n \to (-\infty, \infty]$ satisfies the growth condition

$$\lim_{\|\boldsymbol{x}\|\to\infty} \frac{f(\boldsymbol{x})}{\|\boldsymbol{x}\|} = \infty. \tag{A.61}$$

Proposition A.2 *If the convex function $f : \mathbb{R}^n \to (-\infty, \infty)$ satisfies the growth condition* (A.61)*, then the conjugate function is convex and everywhere finite. If, additionally, the functions f and f^* are both C^1 on $\mathbb{R}^n$, then we have the equality*

$$f^*(\boldsymbol{p}) + f\big(f^*_{\boldsymbol{p}}(\boldsymbol{p}) \big) = \big(\boldsymbol{p}, f^*_{\boldsymbol{p}}(\boldsymbol{p}) \big), \quad \forall \boldsymbol{p} \in \mathbb{R}^n. \tag{A.62}$$

Proof By Proposition A.1, the function f is continuous. If f satisfies (A.61), then we deduce from (A.60) that

$$-\infty < f^*(\boldsymbol{p}) < \infty, \quad \forall \boldsymbol{p} \in \mathbb{R}^n.$$

On the other hand, $f^*(\boldsymbol{p})$ is convex since it is the supremum of a family of convex functions. Since it is everywhere finite, it is continuous by Proposition A.1.

Let us now observe that for any $\boldsymbol{p} \in \mathbb{R}^n$ there exists an $\boldsymbol{x}^* \in \mathbb{R}^n$ such that

$$f^*(\boldsymbol{p}) = (\boldsymbol{p}, \boldsymbol{x}^*) - f(\boldsymbol{x}^*). \tag{A.63}$$

Indeed, there exists a sequence $\{\boldsymbol{x}_\nu\}_{\nu\geq 1} \subset \mathbb{R}^n$ such that

$$(\boldsymbol{p}, \boldsymbol{x}_\nu) - f(\boldsymbol{x}_\nu) \leq f^*(\boldsymbol{p}) \leq (\boldsymbol{p}, \boldsymbol{x}_\nu) - f(\boldsymbol{x}_\nu) + \frac{1}{\nu}, \quad \forall \nu = 1, 2, \dots. \tag{A.64}$$

The growth condition implies that the sequence $\{\boldsymbol{x}_\nu\}_{\nu\geq 1}$ is bounded. The Bolzano–Weierstrass theorem implies that this sequence contains a subsequence $\boldsymbol{x}_{\nu_k}$ that converges to some $\boldsymbol{x}^* \in \mathbb{R}^n$. Passing to the limit along this subsequence in (A.64), we deduce (A.63).

On the other hand, from (A.60) we deduce that

$$f(\boldsymbol{x}^*) \geq (\boldsymbol{x}^*, \boldsymbol{q}) - f^*(\boldsymbol{q}), \quad \forall \boldsymbol{q} \in \mathbb{R}^n.$$

The last inequality, together with (A.63), shows that

$$f(\boldsymbol{x}^*) = \sup\big\{(\boldsymbol{x}^*, \boldsymbol{q}) - f^*(\boldsymbol{q});\ \ \boldsymbol{q} \in \mathbb{R}^n\big\} = (\boldsymbol{x}^*, \boldsymbol{p}) - f^*(\boldsymbol{p}).$$

In other words, $\boldsymbol{p}$ is a maximum point for the function $\Phi(\boldsymbol{q}) = (\boldsymbol{x}^*, \boldsymbol{q}) - f^*(\boldsymbol{q})$ and, according to Fermat's theorem, we must have

$$\Phi_{\boldsymbol{q}}(\boldsymbol{p}) = 0,$$

that is,

$$\boldsymbol{x}^* = f^*_{\boldsymbol{p}}(\boldsymbol{p}).$$

Equality (A.62) now follows by invoking (A.63).

References

1. Arnold, V.I.: Ordinary Differential Equations. Universitext, Springer Verlag (1992)
2. Barbu, V.: Nonlinear Semigroups and Differential Equations in Banach Spaces. Noordhoff, Leyden (1976)
3. Brauer, F., Nohel, J.A.: Qualitative Theory of Ordinary Differential Equations. W.U. Benjamin, New York, Amsterdam (1969)
4. Braun, M.: Differential Equations and Their Applications, 4th edn. 1993. Springer (1978)
5. Clamrogh, N.H.M.: State Models of Dynamical Systems. Springer, New York, Heidelberg, Berlin (1980)
6. Corduneanu, C.: Principles of Differential and Integral Equations. Chelsea Publishing Company, The Bronx New York (1977)
7. Courant, R.: Partial Differential Equations. John Wiley & Sons, New York, London (1962)
8. Crandall, M.G., Lions, P.L.: Viscosity solutions of Hamilton-Jacobi equations. Trans. Amer. Math. Soc. **277**, 1–42 (1983)
9. Halanay, A.: Differential Equations: Stability, Oscillations. Time Logs, Academic Press (1966)
10. Gantmacher, F.R.: Matrix Theory, vol. 1–2. AMS Chelsea Publishing, American Mathematical Society (1987)
11. Hale, J.: Ordinary Differential Equations. John Wiley, Interscience, New York, London, Sidney, Toronto (1969)
12. Kolmogorov, A.N., Fomin, S.V.: Introductory Real Analysis. Dover (1975)
13. Landau, E., Lifschitz, E.: Mechanics. Course of Theoretical physics, vol. 1. Pergamon Press (1960)
14. Lang, S.: Undergraduate Analysis, 2nd edn. Springer (2005)
15. LaSalle, J.P., Lefschetz, S.: Stability by Lyapunov's Direct Method with Applications. Academic Press, New York (1961)
16. Lee, B., Markus, L.: Foundations of Optimal Control. John Wiley & Sons, New York (1967)
17. Pontriagyn, L.S.: Ordinary Differential Equations. Addison-Wesley (1962)
18. Vrabie, I.: Differential Equations. An Introduction to Basic Concepts. Results and Applications. World Scientific, New Jersey, London, Singapore (2011)
19. Whitman, G.B.: Lectures on Wave Propagation. Springer, Tata Institute of Fundamental Research, Berlin, Heidelberg, New York (1979)

V. Barbu, *Differential Equations*, Springer Undergraduate Mathematics Series,
DOI 10.1007/978-3-319-45261-6

Index

V. Barbu, *Differential Equations*, Springer Undergraduate Mathematics Series,
DOI 10.1007/978-3-319-45261-6

Printed in the United States
By Bookmasters